"十四五" 职业教育部委级规划教材

教育部国家职业教育专业教学资源民族文化传承与创新子库

"中国丝绸技艺民族文化传承与创新" 配套双语教材

江苏省高等职业院校高水平专业群 "纺织品检验与贸易" 配套教材

罗织造技艺传承与创新

Inheritance and Innovation of Leno Weaving Technology

赵 兵　牛建涛◎主 编　陈莉霞◎译

柴文波　黄紫娟　黄小萃◎副主编

U0742693

中国纺织出版社有限公司

内容提要 /Summary

本书以中英文对照的形式呈现了中国丝绸织造技艺中罗织造技艺的全貌。全书共五章，包括罗的起源与文化、罗的织造技艺、罗组织的结构、罗的传承与保护、罗的创新与应用。本书图文并茂，语言简洁，通俗易懂，内容兼顾传统与现代、经典与通俗、国内与国外、基础与拓展、实践与应用，体现经典性、实用性和创新性。

本书既可作为高职院校丝绸技术、现代纺织技术、纺织品检验与贸易、纺织品设计等专业的教材，也可作为纺织相关领域工作人员的参考用书。

This book presents a full picture of leno weaving technology, one of Chinese silk weaving, in both English and Chinese. It consists of five chapters, which covers the origin and culture, structure and weaving technology, inheritance and protection as well as innovation and application of leno. The book, which combines tradition and modernity, classic and popularity, domesticity and exoticism, basis and expansion, practice and application, is well illustrated with both texts and pictures and is concise and easy to understand. It is a unity of classic, practicability, and innovation.

This book could serve as a textbook for college students majoring in silk technology, modern textile technology, textile inspection and trade, textile design, etc. It may also serve as a reference book for staff working in textile and art related industries.

图书在版编目（CIP）数据

罗织造技艺传承与创新 =Inheritance and Innovation of Leno Weaving Technology：汉、英 / 赵兵，牛建涛主编；陈莉霞译；柴文波，黄紫娟，黄小萃副主编 . -- 北京：中国纺织出版社有限公司，2024.6

"十四五"职业教育部委级规划教材　教育部国家职业教育专业教学资源民族文化传承与创新子库"中国丝绸技艺民族文化传承与创新"配套双语教材　江苏省高等职业院校高水平专业群"纺织品检验与贸易"配套教材

ISBN 978-7-5229-1097-0

Ⅰ. ①罗… Ⅱ. ①赵… ②牛… ③陈… ④柴… ⑤黄… ⑥黄… Ⅲ. ①绫罗—高等职业教育—教材—汉、英 Ⅳ. ①TS146

中国国家版本馆 CIP 数据核字（2023）第 192805 号

责任编辑：孔会云　　特约编辑：陈彩虹　　责任校对：高　涵
责任印制：王艳丽

中国纺织出版社有限公司出版发行
地址：北京市朝阳区百子湾东里 A407 号楼　邮政编码：100124
销售电话：010—67004422　传真：010—87155801
http://www.c-textilep.com
中国纺织出版社天猫旗舰店
官方微博 http://weibo.com/2119887771
北京通天印刷有限责任公司印刷　各地新华书店经销
2024 年 6 月第 1 版第 1 次印刷
开本：787×1092　1/16　印张：5.5
字数：120 千字　定价：88.00 元

前 言 / **Foreword**

罗织造技艺具有极高的历史文化价值、艺术审美价值、科学认识价值和经济开发价值。然而在现代技术的冲击下，传统罗织造技艺的传承与发展面临巨大的挑战，甚至一度濒临失传。为了更好地在全国甚至全球范围内进行罗织造技艺的传承与创新，苏州经贸职业技术学院组织编写了"中国丝绸技艺民族文化传承与创新"配套双语教材《罗织造技艺传承与创新》。

Leno weaving technology is endowed with high historical and cultural value, artistic aesthetic value, scientific lognition value, and economic development value. However, the impact of modern technology poses a huge challenge for the traditional leno weaving technology, which was almost at the edge of extinction. In order to better inherit and protect leno weaving technology nationwide and worldwide, Suzhou Institute of Trade & Commerce was assembled to compile the bilingual textbook *Inheritance and Innovation of Leno Weaving Technology*, which is one of the series among the "National Culture Inheritance and Innovation of Chinese Silk technology".

本书内容共分为五章，第一章概述罗的起源与文化，由苏州经贸职业技术学院柴文波编写；第二章阐述罗的织造技艺，由苏州经贸职业技术学院牛建涛编写；第三章介绍罗组织的结构，由苏州经贸职业技术学院黄紫娟、苏州市职业大学黄小莘共同编写；第四章论述罗的传承与保护，由苏州经贸职业技术学院赵兵编写；第五章探讨罗的创新与应用，由苏州经贸职业技术学院周青奇编写。全书由赵兵统稿，全书英文由西安工程大学陈莉霞团队翻译。

The book consists of five chapters. The first chapter, written by Chai Wenbo at Suzhou Institute of Trade & Commerce, introduces the origin and culture of leno. The second chapter, written by Niu Jiantao at Suzhou Institute of Trade & Commerce, describes the weaving technology of leno. The third chapter, co-written by Huang Zijuan at Suzhou Institute of Trade & Commerce and Huang Xiaocui at Suzhou Vocational University focuses on the structure of leno. The fourth chapter, written by Zhao Bing at Suzhou Institute of Trade & Commerce, is about the inheritance and protection of leno. The fifth chapter, written by Zhou Qingqi at Suzhou Institute of Trade & Commerce, investigates the innovation and application of leno. The final book is compiled and edited by Zhao Bing. The English version is translated by Chen Lixia and her team at Xi'an

Polytechnic University.

本书在编写过程中，得到苏州经贸职业技术学院的领导、老师和苏州工业园区家明织造坊周家明的大力支持与帮助，在此表示衷心的感谢。

Sincere gratitude should be expressed to the leaders and teachers at Suzhou Institute of Trade & Commerce and Zhou Jiaming from Jiaming Weaving Workshop in Suzhou Industrial Park, who provided great support and assistance to the compilation of the book.

限于编者的能力和水平，书中不妥之处在所难免，敬请广大读者提出宝贵意见，以便今后进一步修订和完善，使之不断进步。

Due to the limited knowledge, there might be some mistakes and flaws in this book. Your comments and suggestions would be greatly valued and appreciated to help us improve in the revision of the book.

编者
2023 年 1 月

翻译前言
Translation Preface

　　《罗织造技艺传承与创新》是教育部国家职业教育专业教学资源民族文化传承与创新子库"中国丝绸技艺民族文化传承与创新"配套双语教材中的一种，也可供对纺织服装感兴趣的人士及纺织服装行业从业者学习参考。本书英译部分旨在引发国内外纺织服装领域从业者的兴趣，为中国纺织服装业的国际拓展发挥作用。

　　Inheritance and Innovation of Leno Weaving Technology, is one of the Bilingual Teaching Materials "Inheritance and Innovation of Chinese Silk Skills and National Culture", a sub-library of national vocational education professional teaching resources of the Ministry of Education. It can also serve as a reference for the readers who are interested in textiles and clothing as well as practitioners in the textile and clothing industry. The English version of the course will be of interest to many practitioners in the industry at home and abroad. It will play an important role in the international expansion of the Chinese textile and clothing industry.

　　本书由西安工程大学陈莉霞副教授及其团队翻译。全书共分五章，第一章由张睿、毛若函翻译，第二章由杨悦翻译，第三章由孙立丛翻译，第四章由刘雨青翻译，第五章由刘凤琴翻译。马淑芬协助校对了这五章的英文版。西安工程大学胡伟华教授审读了所有翻译文稿。

　　It was translated by Associate Professor Chen Lixia and her team at Xi'an Polytechnic University. The book consists of five chapters, with Zhang Rui and Mao Ruohan undertook the translation of the first chapter, Yang Yue the second chapter, Sun Licong the third chapter, Liu Yuqing the fourth chapter and Liu Fengqin the fifth chapter respectively. Ma Shufen helped with the proofreading of the English version of the five chapters. Professor Hu Weihua from Xi'an Polytechnic University profread the whole translation.

译者

2023年8月

教学内容及课时安排

章	课时	节	课程内容
第一章	10课时		第一章　罗的起源与文化
		一	第一节　罗的简介
		二	第二节　商周时期的罗
		三	第三节　春秋战国时期的罗
		四	第四节　两汉时期的罗
		五	第五节　隋唐时期的罗
		六	第六节　宋元时期的罗
		七	第七节　明清时期的罗
第二章	12课时		第二章　罗的织造技艺
		一	第一节　罗织造技艺的演变
		二	第二节　罗生产织造的工序
第三章	14课时		第三章　罗组织的结构
		一	第一节　罗组织的概念和形成过程
		二	第二节　罗组织的上机
		三	第三节　罗组织的应用实例
第四章	12课时		第四章　罗的传承与保护
		一	第一节　罗的传承现状
		二	第二节　罗织造技艺传承人
		三	第三节　罗传承与发展面临的困境
		四	第四节　罗的保护、传承与发展建议
第五章	6课时		第五章　罗的创新与应用
		一	第一节　苏州地域文化及其视觉化提炼
		二	第二节　苏州地域文化在罗织物设计中的创新应用
		三	第三节　桃花坞年画元素在罗织物设计中的创新应用
		四	第四节　昆曲元素在罗织物设计中的创新应用

注　各院校可根据自身的教学特点和教学计划对课程时数进行调整。

Teaching Content and Class Hours

Chapter/ Class Hour	Course Type/Class Hour	Class Period	Contents
Chapter 1	10 Class Hours		**1. The Origin and Culture of Leno**
		1	1.1 The Brief Introduction to Leno
		2	1.2 Leno in the Shang and Zhou Dynasties
		3	1.3 Leno in the Spring and Autumn Period and the Warring States Period
		4	1.4 Leno in the Han Dynasty
		5	1.5 Leno in the Sui and Tang Dynasties
		6	1.6 Leno in the Song and Yuan Dynasties
		7	1.7 Leno in the Ming and Qing Dynasties
Chapter 2	12 Class Hours		**2. The Weaving Technology of Leno**
		1	2.1 The Evolution of Leno Weaving Technology
		2	2.2 The Manufacturing and Weaving Process of Leno
Chapter 3	14 Class Hours		**3. The Structure of Leno**
		1	3.1 The Concept and Formation Process of Leno Weaves
		2	3.2 The Looming Plans of Leno Weaves
		3	3.3 The Application Examples of Leno Weaves
Chapter 4	12 Class Hours		**4. The Inheritance and Protection of Leno**
		1	4.1 The Current Situation of Leno Inheritance
		2	4.2 The Inheritors of Leno Weaving Technique
		3	4.3 The Difficulties in Leno Inheritance and Development
		4	4.4 Proposals for Leno Protection, Inheritance and Development
Chapter 5	6 Class Hours		**5. The Innovation and Application of Leno**
		1	5.1 An Overview and Visual Refinement of Suzhou Regional Culture
		2	5.2 The Innovative Application of Suzhou Regional Culture in Leno Fabric Design
		3	5.3 The Innovative Application of Taohuawu New Year Pictures Elements in Leno Fabric Design
		4	5.4 The Innovative Application of Kunqu Opera Elements in Leno Fabric Design

Note   Class hours can be adjusted according to the teaching curriculum and arrangement of individual schools.

目 录/Contents

第一章　罗的起源与文化/The Origin and Culture of Leno ··································· 1

第一节　罗的简介/The Brief Introduction to Leno ······························· 1

第二节　商周时期的罗/Leno in the Shang and Zhou Dynasties ················· 6

第三节　春秋战国时期的罗/Leno in the Spring and Autumn Period
and the Warring States Period ································· 7

第四节　两汉时期的罗/Leno in the Han Dynasty ······························ 8

第五节　隋唐时期的罗/Leno in the Sui and Tang Dynasties ···················· 10

第六节　宋元时期的罗/Leno in the Song and Yuan Dynasties ·················· 12

第七节　明清时期的罗/Leno in the Ming and Qing Dynasties ·················· 14

第二章　罗的织造技艺/The Weaving Technology of Leno ······················· 17

第一节　罗织造技艺的演变/The Evolution of Leno Weaving Technology ········ 17

一、史前的绞经与绞编织造技艺/The Twist Knitting Weaving Technology of
Strand Textiles in Prehistory ································· 17

二、商周时期的绞经织造技艺/The Weaving Technology of Strand Textiles in
the Shang and Zhou Dynasties ································ 19

三、战国至秦汉时期的绞经织造技艺/The Weaving Technology of Strand Textiles
from the Warring States Period to the Qin and Han Dynasties ············· 21

四、隋唐时期的罗织造技艺/The Weaving Technology of Leno Fabrics in the Sui
and Tang Dynasties ·· 22

五、元时期的罗织造技艺/The Weaving Technology of Leno Fabrics in the Yuan
Dynasty ·· 23

六、明清时期的罗织造技艺/The Weaving Technology of Leno Fabrics in the
Ming and Qing Dynasties ·································· 24

第二节　罗生产织造的工序/The Manufacturing and Weaving Process of Leno ······· 25

一、原料准备 /Raw Material Preparation ···························· 25

二、织造准备/Weaving Preparation ·· 26

三、上机织造/Weaving Process ·· 28

四、后整理/Post-finishing ·· 29

五、现代丝绸罗的加工织造/Weaving Process of Modern Silk Leno ········· 29

第三章　罗组织的结构/The Structure of Leno ································· 31

第一节　罗组织的概念和形成过程/The Concept and Formation Process of Leno

Weaves ··· 31

一、罗组织的基本概念/The Basic Concepts of Leno Weaves ·············· 31

二、罗组织的形成/The Formation of Leno Weaves ························ 33

第二节　罗组织的上机/The Looming Plans of Leno Weaves ·················· 40

一、上机图的绘制/Graphic Design for the Looming Plans of Leno Weaves ········· 40

二、上机要点/The Key Point of Weaving ································· 43

第三节　罗组织的应用实例/The Application Examples of Leno Weaves ········· 43

第四章　罗的传承与保护/The Inheritance and Protection of Leno ············ 46

第一节　罗的传承现状/The Current Situation of Leno Inheritance ··········· 46

一、杭罗传承现状/Current Situation of Hang Leno Inheritance ··········· 47

二、吴罗（四经绞罗）传承现状/The Current Situation of Wu Leno（Four-warp

Twisted Leno）Inheritance ·· 49

第二节　罗织造技艺传承人/The Inheritors of Leno Weaving Technique ········· 51

一、杭罗织造技艺传承人/Inheritors of Hang Leno Weaving Technique ········· 51

二、吴罗织造技艺传承人/Inheritors of Wu Leno Weaving Technique ········· 52

第三节　罗传承与发展面临的困境/The Difficulties in Leno Inheritance and

Development ·· 55

一、代表性传承人老龄化严重，缺乏接班人/Aging of Representative Inheritors

and Lack of Successors ·· 55

二、缺乏色彩、图案设计人员，制约罗新产品开发/Lack of Color and Pattern

Designers, Restricting the Development of New Leno Products ············ 56

三、织造工艺复杂，效率低，生产成本高/Complex Weaving Process,

Low-efficiency and High Production Costs ································ 56

四、产品出口量减少，国内消费趋于小众市场/Export Trades Decline while

Domestic Market Niche ·· 57

第四节　罗的保护、传承与发展建议/Proposals for Leno Protection, Inheritance and Development ··· 57

一、建立罗织造技艺的非遗生态圈/Establishing an Intangible Cultural Heritage Ecosystem for Leno Weaving Technique ······································· 57

二、完善罗织造技艺的保护传承发展平台/Improving the Protection and Development Platform of Leno Weaving Technique ························· 60

三、丰富罗织造技艺的宣传推广形式/Enriching Forms of Publicity and Promotion of Leno Weaving Technique ··· 63

四、在地方高校中传承丝绸非遗技艺和文化，充实传承人队伍/Passing on Silk Intangible Cultural Technique and Culture in Local Institutions of Higher Learning to Enrich the Team of Inheritors ································· 64

五、搭建产学研合作平台，推动罗产品的开发与创新设计/Building a Platform for Industry–university–research Cooperation to Promote the Development and Innovative Design of Leno Products ································· 64

六、推动罗织造技艺的工艺研发/Promoting Research and Development in Leno Weaving Technique ··· 65

第五章　罗的创新与应用/The Innovation and Application of Leno ··········· 67

第一节　苏州地域文化及其视觉化提炼/An Overview and Visual Refinement of Suzhou Regional Culture ··· 67

第二节　苏州地域文化在罗织物设计中的创新应用/The Innovative Application of Suzhou Regional Culture in Leno Fabric Design ····················· 69

第三节　桃花坞年画元素在罗织物设计中的创新应用/The Innovative Application of Taohuawu New Year Pictures Elements in Leno Fabric Design ····· 70

第四节　昆曲元素在罗织物设计中的创新应用/The Innovative Application of Kunqu Opera Elements in Leno Fabric Design ····························· 72

参考文献/References ·· 75

○ 第一章

罗的起源与文化
The Origin and Culture of Leno

第一节　罗的简介/The Brief Introduction to Leno

在古往今来众多丝绸产品中，罗具有以下两大特征：一是高档，"遍身罗绮者，不是养蚕人"，描述的就是着罗衣者的富贵；二是轻薄，在影视作品中，无论是环肥燕瘦的人间女子，还是不食人间烟火的仙女，最切合的穿着总是一袭罗衣。不同于锦衣缎袍的华贵、刺绣缂丝的富丽，罗衣的美，仿如翩翩公子居于草堂，绝代佳人在水一方，别有一种风流袅娜的韵味。

Among many silk products through the ages, leno has two characteristics. Firstly, it is a top-end product. As described in a Chinese Song poem, "people wearing silk clothes were not silk growers" implied that this person was rich. Secondly, it is light and thin. In the film and TV works, whether she is an ordinary mortal attractive in her own way, or a goddess who is pure and saintly, always wears leno clothes. Different from the luxury of brocade and satin clothes and the richness of embroidery and kossu, the charm of leno clothes, as a handsome young man living in a simple cottage, a peerless beauty on the waterside, has a charm of elegance and romance.

罗，最初不是指丝绸织物，而是指捕鸟的网。东汉许慎在《说文解字》中将"罗"解释为"以丝罟鸟"，故其字意为"网""维"。《尔雅·释器》称："鸟罟谓之罗。"《诗经·王风》中有"雉离于罗"，我们今天所说的"网罗""天罗地网"等，其中的"罗"都是"网""维"的意思。《说文解字》以"古者芒氏初作罗"，《周礼·夏官》中有"罗氏"，"罗"是指专门从事捕鸟活动的人。而最初的纺织品，因为没有机织工具，只能手工编织，"手经指挂，其成犹网罗"，所以经纬密度粗疏得像捕鸟的网，故称为"罗"。随着纺捻工具的发展，经纬密度越来越细密，织物变得平整且可织出花纹，品种也越来越多。

Leno did not originally refer to silk fabrics, but the net for catching birds. Xu Shen in the Eastern Han Dynasty explained "Leno" as "catching birds with woven nets" in *The Origin of Chinese Characters*, meaning "net" or "dimension". *Erya of Explanation of Tools* (*Erya*, the first dictionary in China) also said "tools used for catching birds can be called Leno". *Wangfeng* of the Book of songs "preys cannot flee from the net", as well as "Wang Luo (nets and snares, meaning "traps" or "searching from all aspects")" and "Tian Luo Di Wang (nets above and snares below)" today and so on share the meaning of "net". According to *The Origin of Chinese Characters,* leno was first created by one of the Chinese ancestors, Mang. In *Rites of Zhou* of *Xia offices,* "leno" referred to the person who specialized in bird catching. Since there were no weaving tools at that time, the original textile could only be woven manually. That's why people say "nets and snares finally made with bare hands". The warp and weft yarn density was as low as a bird's net and could only be used for catching birds. With the development of weaving and spinning tools, the count of fabrics gets finer, the fabrics become flat and can be woven with patterns and with more varieties.

现行织物组织学一般认为凡经线发生相绞、纬线平行交织的织物均可称为纱罗织物，其组织即为纱罗组织。"罗"与"纱"都质地轻薄、表面有孔眼，两者常常相提并论，合称"纱罗"，如图1-1所示。罗其实是可以织得较为厚实的，但在人们的印象中，似乎轻盈的罗才是真正的罗，与纱相似。"罗"与"纱"常常被混为一谈，但从织物结构上说，两者还是有区别的。在汉唐时期，"纱"是经纬线垂直交织的平纹结构，形成的孔眼是平直的；而"罗"的经线是两两交织相绞着与纬线交织，绞经结构形成的孔眼相对固定，因而又有"方孔曰纱，椒孔曰罗"的说法。明代以后，又将表面有横向或纵向平纹条纹的绞经织物称为"罗"，而将绞经形成的孔眼连成一片的称为"纱"。罗在我国古代的用途非常广泛，既可以用作服饰，又可以作为各种覆盖、装饰、包装的用料，还可以用作帷帐和糊窗的材料，是中国古代一种生产量很大、应用很普遍的丝织品。

The current fabric texture studies generally believe that any fabric in which the warp threads are entangled with each other and the weft threads are intertwined in parallel can be called gauze (yarn and leno fabrics), and its texture is the leno texture. Leno and yarn are often mentioned in the same breath and collectively referred to as gauze, as shown in Figure 1-1, for both of them sharing the features of lightness and thinness as well as having eyelets on surfaces. In fact, leno can be woven thicker, but people always have the impression that the real leno fabrics should be light and they cannot quite tell the difference between leno and yarn. Leno and yarn are often regarded as the some thing, but in terms of fabric structure, there are still differences between them. In the Han and Tang Dynasties, the warp and weft threads of "yarn" fabrics were vertically intertwined, with a plain structure, and the eyelets formed were straight; while the warp threads of "leno" fabrics were intertwined with the weft threads in pairs, and the eyelets formed by the strand warp structure were relatively fixed. That's why there was also a saying that "yarn with square eyelets, but leno with chili-seed-shape eyelets". After the Ming Dynasty, the strand warp fabrics with horizontal or vertical plain stripes on surfaces were called "leno fabrics", and what connect the eyelets formed by strand warps was called "yarn". Leno fabrics were of extensive use in ancient China. They could be used not only for clothes, but also for various covering, decoration and packaging materials as well as curtains and window pasting materials. They were silk fabrics with large production volume and common application in ancient China.

从目前的考古发现来看，最早出土的丝织品属于新石器时代，自商代伊始直至唐代，罗织物逐步发展成熟，这段时期罗织物的主要组织类型是链式罗，地经和绞经之比为1：1，地经和绞经相间排列，但一根绞经可以和相邻的两根地经起绞的罗织物被称作四经绞罗，是链式罗的一种。

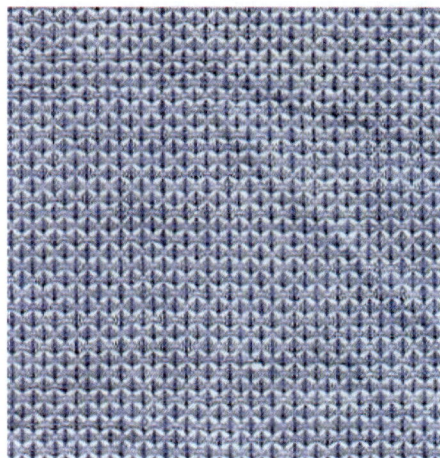

图1-1　纱罗织物
Gauze

According to current archaeological discoveries, the earliest among the unearthed silk fabrics can be traced back to the Neolithic Age. Leno fabrics had gradually developed from the beginning of the Shang Dynasty to the Tang Dynasty, during which the main texture type of leno fabrics was chain leno, the ratio of their strand warps and Strand warps was 1：1, and the strand warps and strand warps were arranged alternately. But the leno fabrics formed by one strand warp twisting two adjacent strand warps were called four-warp crossing leno, which was a kind of chain leno fabrics.

距今6000年左右的西安半坡仰韶文化遗址出土的陶器上，有很多织物印痕，大多为平纹织物印痕，但也有少量绞罗织物印痕（图1-2）。自进入文明社会以来，关于罗作为达官贵人的服饰和家居用料的记载更是不胜枚举。先秦两汉时期，北方中原地区与南方蜀（今四川）、楚（今湖南和湖北）之地的丝绸业特别发达。中原的丝织品生产中心在今河南省与山东省，出产罗、纨、绮、锦、绣等纤丽之物。南方的丝织品生产中心在四川成都，主要出产有名的蜀锦。文献中对荆楚地区（今湖北省及周边地区）丝绸业的记载不多，但战国至汉代的楚墓出土了大量精美的丝织品，如1982年发现的湖北江陵马山一号战国楚墓和1972年发现的湖南长沙马王堆一号西汉墓，均出土了罗织物，包括素罗和花罗，体现了极高的工艺技术水平。

On the potteries unearthed from the Yangshao Cultural Relics in Banpo, Xi'an with a history of about 6,000 years, there are many fabric imprints, of which most are plain fabric imprints and a few are crossing leno fabric imprints（Figure 1-2）. Since entering a civilized society, leno fabrics, as clothes and household materials for dignitaries, appeared frequently in literature. In the pre-Qin and Han Dynasties, silk fabric production was particularly developed in the Central Plains in China's northern area and the Shu（now Sichuan）and Chu（now Hunan and Hubei）regions in China's southern area. The production centers of silk fabrics in the Central Plains were in today's Henan and Shandong Provinces, producing delicate and beautiful leno, Wan, Qi, brocade and embroider fabrics（all refer to fine silk fabrics）. The silk production center in China's southern area was Chengdu, Sichuan which mainly produced the famous Shu brocade . There are not many records of the silk industries in the Jingchu area（now Hubei Province and areas around）, but a large number of

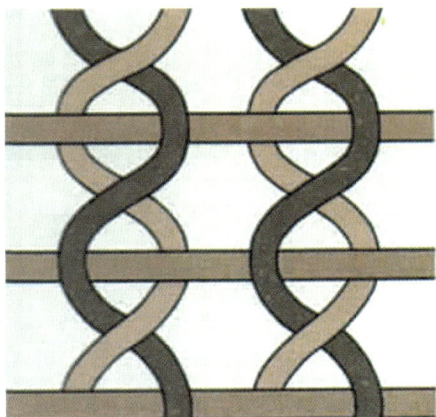

图1-2　绞罗织物印痕
Crossing Leno Fabric Imprints

exquisite silk fabrics have been unearthed from the Chu tombs of the Warring States Period to the Han Dynasty. From the Jiangling Mashan No. 1 Warring States tomb excavated in 1982 and the No. 1 Western Han Dynasty tomb discovered in 1972, leno fabrics, including plain leno and figured leno fabrics, were unearthed, reflecting an extremely high level of craftsmanship.

产于浙江杭州的罗，称杭罗或浙罗。杭罗与苏缎、云锦同为中国华东地区的三大丝绸名产。"杭罗织造技艺"作为中国蚕桑丝织技艺中的重要代表性项目，已于2009年9月30日经联合国教科文组织批准列入《人类非物质文化遗产代表作名录》。

Leno, produced in Hangzhou, Zhejiang province, is called Hang Leno or Zhe Leno. Hang Leno, along with Su Satin and Yun Brocade, are the three famous silk products in East China. The Hang Leno Weaving Technique, as an important representative project of Chinese sericulture and silk weaving techniques, was entered in the *Representative List of the Intangible Cultural Heritage of Humanity*, which was approved by UNESCO on September 30, 2009.

杭罗由纯桑蚕丝以平纹和纱罗组织联合构成，有横罗和直罗两种，具有等距规律的直条形或横条形纱孔，孔眼清晰，穿着舒适凉快。在杭罗上刺绣的方法称为戳纱或者挑罗。

Hang Leno is composed of pure mulberry silk with a combination of plain weave and gauze tissue. It has two types: horizontal weave and straight weave, with equidistant regular straight or horizontal gauze holes, clear eyelets, comfortable and cool to wear. The method of embroidery on Hang Leno is called poking yarn or picking silk.

杭罗历史悠久，早在宋代地方志中就有记载，至清代成为杭州丝绸的著名品种。生产杭罗的织机几经变革，但其生产流程中仍保持着大量精细缜密的手工技艺，对生产者技能的要求极高。原料丝进厂后，必须经过严格检验、筛选，历经浸泡、晾干、翻丝、纤经、摇纡等系列工艺，然后才能上机织造。织成的粗坯还要经过精练、染色等工序，才能成为精致的杭罗。

Hang Leno has a long history and was recorded in local chronicles as early as the Song Dynasty. By the Qing Dynasty, it had become a famous variety of silk in Hangzhou. The weaving machines used in the production of Hang Leno have undergone several changes, but their production process still maintains a large amount of meticulous manual skills, requiring extremely high levels of skill from producers. After the raw silk enters the factory, it must be strictly inspected and screened, and undergo a series of processes such as soaking, drying, flipping, warp knitting, and shaking before being woven on the machine. The woven rough fabric needs to go through processes such as refining and dyeing in order to become exquisite Hang Leno.

区别于杭罗或浙罗，苏州地区的丝绸生产者习惯上把当地生产的真丝罗称为苏罗或吴罗。苏州地区的丝绸业历史悠久，可以追溯到新石器时期。该地区也曾出土过四千多年前的丝线、丝带、绢片等实物，特别是其中的绢片，其经纬交织的细密与平整，使人们猜想当时的丝织技术已达到了相当成熟的程度。先秦时期，苏南地区先后为越、吴、楚三国所属，秦汉时期中原文化对苏南地区的影响逐渐增大。汉末三国时期，孙吴起于苏南地区，与魏、蜀成三国鼎立之势。至东晋南朝时期，北方中原地区战乱，大量衣冠豪族东渡，苏南地区经济文化有了更大的发展，文献上对该地丝绸业的记载也开始增多。隋朝时期开始挖大运河，沟通了苏南地区与中原地区的交通大动脉。到了唐朝时期，特别是安史之乱后，由于中原地区战乱，江南地区开始摆脱不重要的边地身份，逐渐走向历史舞台的中央，丝绸业也迅速发展起来。南宋时期，全国的政治、经济与文化中心南迁至江南地区，逐步奠定了江南地区丝绸生产中心的地位。到明清时期，江南织造府成立，江南地区已经成为名副其实的"丝绸之府"。

In order to distinguish it from Hang Leno or Zhe Leno, silk producers in Suzhou region commonly refer to locally produced silk as Su Leno or Wu Leno. The development of the silk industry in Suzhou started very early and can be dated back to the Neolithic era. The silk threads, ribbons, and silk slices with a history of over 4,000 years were unearthed here. In particular, the silk slices, whose warp and weft interweaving is fine and smooth, make people guess that the silk weaving technology at that time had reached a quite mature degree. In the Pre–Qin Dynasty, the southern Jiangsu region was dominated by Yue, Wu, and Chu vassal states, so the Central Plains culture exerted an increasing influence on the southern Jiangsu region in the Qin and Han Dynasties. During the Three Kingdoms Period in the late Han Dynasty, Sun Family founded the Wu Kingdom in the southern Jiangsu region which rivaled with Wei and Shu Kingdoms. During the Eastern Jin Dynasty and the Southern Dynasties, there were wars in the Central Plains in the north, and a large number of rich and noble families migrated eastward. So the economy and culture of the southern Jiangsu region gained a greater development, and the literature about the silk industry began to increase. In the Sui Dynasty, the Grand Canal was excavated, connecting the traffic artery of the southern Jiangsu and Central Plains. In the Tang Dynasty, especially after the An–Shi Rebellion, constant wars in the Central Plains pushed the Jiangnan region towards the center of the historical stage, and the silk industry developed rapidly at that time. During the Southern Song Dynasty, the national political, economic and cultural centers moved south to the Jiangnan region, gradually laid foundation for the Jiangnan region as a silk production center. By the Ming and Qing Dynasties, the "Jiangnan Weaving House" was established. Till that time, the Jiangnan area has become a veritable "hometown of silk fabrics".

从《吴县志》的记载来看，吴地的丝织品生产历史久远，从三国时期已经开始对外丝绸贸易，北宋至明初官营丝织业发展繁荣，明清时期达到鼎盛，文献对此有详细的描述："明嘉靖至万历年间，丝织形成独立而庞大的手工业行业，城厢成为全国丝绸生产的重要基地，织工达数千人。清康熙时，官营织机达800张，机匠2230名。乾隆年间，'比户习织专其业者不啻万家'，机户遍布城区及东北的唯亭、蠡口和境西的香山、光福等，成立'吴郡机业公所''云锦公所'。道光廿年（1840年）织机达1.2万台……至光绪廿年（1894年），织机增至15万台，织工3万余人。"

According to the *Local Records of Wu County*, silk fabric production in Wu region has a long history. Foreign silk trade has been going on since the Three Kingdoms Period, the official silk weaving industry developed and boomed from the Northern Song Dynasty to the early Ming Dynasty and reached its peak in the Ming and Qing Dynasties. According to a historical record, from the reign of the Emperor Jiajing to the reign of the Emperor Wanli in the Ming Dynasty, silk weaving had formed an independent and huge handicraft industry, and Chengxiang had become a nationally important silk production base with thousands of weavers. During the reign of the Emperor Kangxi of the Qing Dynasty, there were 800 official looms and 2,230 weavers. During the reign of the Emperor Qianlong, almost all families were learning weaving, leading to more than ten thousand people specializing in that. And the households with looms were all over the urban area and Weiting, Likou in the northeast, Xiangshan and Guangfu in the west of Wu County, etc., and the "Wu County Weaving Industry Office" as well as the "Yun Brocade Office" was established. In the 20th year of the Daoguang Emperor's reign（1840）, the number of looms reached 12,000, and in the 20th year of the Guangxu Emperor's reign（1894）, the number of looms increased to 150,000, and there were more than 30,000 weavers.

第二节　商周时期的罗/Leno in the Shang and Zhou Dynasties

　　我国出土的最早的绞编织物是距今约6000年前的江苏草鞋山新石器时期遗址中的三块野生葛绞编织物，这是我国目前出土的最早的纺织实物，充分证明了绞编织物及其制作技术至少在人类新石器时期已经成熟和盛行，由于其所具有的罗的互绞的特征明确，因此认为这是人类现有已考证的最早的罗织物实物。

The earliest twisted woven fabric unearthed in China is three wild kudzu twisted woven fabrics from the Neolithic site in Caoxie Mountain, Jiangsu Province, about 6,000 years ago. This is the earliest textile object unearthed in China, fully proving that twisted woven fabrics and their production technology were mature and popular at least in the Neolithic period of human beings. Due to their clear characteristics of intertwining of the Lenos, it is believed that this is the earliest verified woven fabric object in human existence.

　　河南荥阳青台村仰韶文化遗址出土了浅绛色的罗织物，这是我国出土的最早的采用桑蚕丝织造的罗织物，同时也是我国目前发现最早的具有色泽的丝织物。

The Yangshao Cultural Site in Qingtai Village, Xingyang, Henan Province unearthed a light reddish colored silk fabric, which is the earliest silk fabric woven from mulberry silk unearthed in China and also the earliest silk fabric with color discovered in China.

商周时期，我国的蚕桑丝织技术有了很大的进步。由于这一时期的君王是巫觋文化的崇拜者，十分迷信，丝织中的精品——罗，常被用来包覆王宫的祭祀礼器。因此有关这一时期罗织物的考古发现，基本上都是在青铜器、玉器或陶器上留下的印记。从出土的罗织物痕迹来看，这一时期已经出现了简单的罗织机，可能当时的普通织机只需增加一两片起绞的综片，便能够织造简单的罗织物。

During the Shang and Zhou Dynasties, China's sericulture and silk weaving technology had made significant progress. Due to the fact that the monarchs of this period were worshippers of shamanism culture and were highly superstitious, the silk weaving of Leno was often used to wrap the sacrificial vessels in the palace. Therefore, archaeological discoveries about woven fabrics during this period were mostly imprinted on bronze, jade, or pottery. From the traces of unearthed woven fabrics, it can be seen that simple woven machines had already appeared during this period. Perhaps it was on ordinary looms at that time that adding one or two twisted warp pieces could weave simple woven fabrics.

第三节　春秋战国时期的罗/Leno in the Spring and Autumn Period and the Warring States Period

春秋战国时期，丝绸业无论是从生产的工艺技术的提高和织造机器的创新，还是从丝绸产品的生产数量的大幅提高来说，进步都是十分显著的。其中，当时楚文化的发源地，如今的两湖地区十分注重发展丝织技术。《管子·小匡》中记载楚国："贡丝于周室"，说明楚国在当时已经具有相当大的丝绸生产规模。因此在这一时期的考古中，出土了大量的纺织品，有些甚至保存得十分完好。

During the Spring and Autumn Period and the Warring States Period, the silk industry made significant progress in terms of improving production techniques and innovating weaving machines, as well as significantly increasing the production quantity of silk products. Among them, the birthplace of Chu culture at that time and the present-day Two Lakes region attach great importance to the development of silk weaving technology. In the *Guanzi Xiaokuang*, it is recorded that the State of Chu "tribute silk was given to the Zhou Dynasty," indicating that the State of Chu already had a considerable scale of silk production at that time. Therefore, in the archaeological discoveries of this period, a large number of textiles were unearthed, some of which were even well preserved.

湖北江陵马山一号墓出土了大量的丝织品，几乎包括了我国先秦丝织品的所有品种，被誉为战国时期的"丝绸宝库"。其中，该墓葬中出土了一件龙凤虎纹绣罗禅衣，是唯一以罗织物作为绣地的绣品。该罗织物为素罗，以龙、凤、虎构成一单位花纹。罗织物孔眼均匀，有半透明的状态，使用这样的面料作为绣地，使整个图案看上去更加立体、形象，具有更好的反衬作用。

A large number of silk fabrics, including almost all varieties of pre-Qin silk fabrics in China, were unearthed from Tomb No.1 of Mashan in Jiangling, Hubei, and are known as the "silk treasure trove" of the Warring States period. Among them, a dragon, phoenix, and tiger embroidered Leno Chan robe was unearthed from the tomb, which is the only embroidery piece that uses Leno fabric as the embroidery site. The fabric is made of silk, with a unit pattern consisting of dragons, phoenixes, and tigers. The eyelets of the woven fabric are uniform and have a semi-transparent state. Using this fabric as the embroidery floor makes the entire pattern look more three-dimensional and vivid, with better contrast effect.

第四节　两汉时期的罗/Leno in the Han Dynasty

两汉时期，我国的丝绸生产规模和生产技术迅速发展，人们已经开始采用提花技术织造具有简单图案的花罗织物。文献中的"纱""罗"织物出现的频率大大提高，考古发现的罗织物数量也大大增多。

During the Han Dynasty, China's silk production scale and technology developed rapidly, and people began to use jacquard technology to weave floral fabrics with simple patterns. The frequency of the appearance of "yarn" and "silk" fabrics in literature greatly increased, and the number of silk fabrics discovered in archaeology has also greatly increased.

湖南长沙马王堆一号汉墓中出土完好保存的丝织品和服饰约一百余件，其中罗织物共有10件，经线密度大约为100~120根/cm，纬线密度大约为35~40根/cm。从组织结构来看，罗织物不仅包括早在商周时期就出现的二经相绞、四经相绞的素罗，还出现了具有花纹图案的花罗织物。这些花罗织物也被用作成品，如手套、香囊、夹袍和帷幔等，还有一些作为绣品的地面料。

More than 100 well preserved silk fabrics and clothing were unearthed from the No.1 Han Tomb in Mawangdui, Changsha, Hunan. Among them, there are a total of 10 woven fabrics, with a warp density of about 100–120 pieces/cm and a weft density of about 35–40 pieces/cm. From the perspective of organizational structure, Leno fabrics not only include Su Leno, which appeared as early as the Shang and Zhou dynasties, with two warp and four warp intertwined, but also include Hua Leno fabrics with patterned patterns. These floral fabrics are also used as finished products, such as gloves, sachets, jackets, and curtains, as well as some floor fabrics for embroidery.

其中出土的菱纹罗是距今保存最早最完整的罗织物。织物上有明暗相间的两种花纹，菱形花纹排列紧凑、上下对称、精致秀丽。组织结构上，罗织物是以四经绞罗为地，二经绞罗为花，两种组织复合形成的罗组织，如图1-3所示。而且这些织物在织成后进行染色，有皂、烟、朱红等色彩，是当时十分高级的丝绸面料。可以看出，从汉代开始，已经有结构复杂且有组织形式复合的罗织物，即已经开始将多个单一的罗组织进行复合形成的花罗组织。

The diamond patterned Leno fabric unearthed among them is the earliest and most complete preserved Leno fabric to this day. There are two patterns of light and dark on the fabric, with diamond patterns arranged tightly, symmetrically up and down, and exquisite and beautiful. In terms of organizational structure, Leno fabric is a Leno structure formed by the combination of four warp twisted Leno as the ground and two warp twisted Leno as the flower, as shown in Figure 1-3. And these fabrics were dyed after weaving, with colors such as soap, smoke, and vermilion, making them very advanced silk fabrics at that time. It can be seen that since the Han Dynasty, there have been complex and structurally complex woven fabrics, that is, flower silk fabrics formed by combining multiple single woven fabrics.

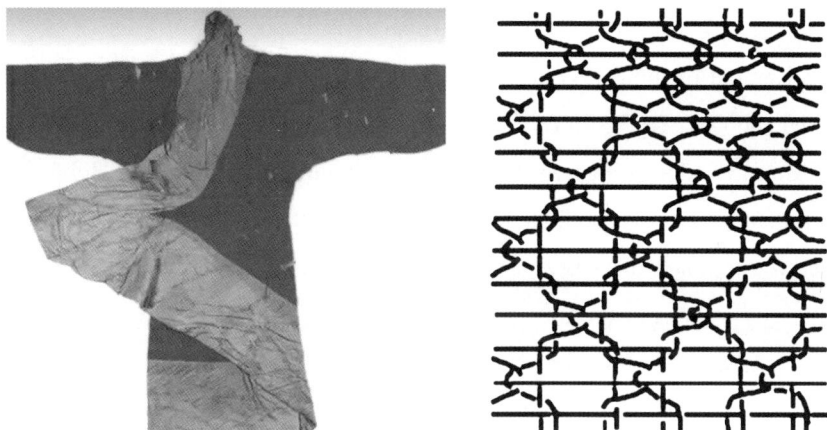

(a) 朱红菱纹罗丝锦袍/Vermilion Diamond Leno Silk Robe   (b) 罗组织/Weave Structure of Leno

图1-3 湖南长沙马王堆出土的菱纹罗丝织物及其织物组织图
Diamond Leno and Its Weave Structure Excavated in the Tomb No.1 of the Han Dynasty at Mawangdui in Changsha City, Hunan Province

除此之外，在吐鲁番盆地的天山阿拉沟的30号墓葬中，也发现了菱纹链式罗的痕迹。新疆民丰出土的花罗织物也是由这样的结构织成，经密66根/cm，纬密26根/cm。蒙古诺因乌拉东汉墓和朝鲜东汉王盱墓出土的菱纹罗也都与此罗织物相同。这些地方都并非古代要道，但同样出土了中原的罗织物，说明这一时期的罗织物已经通过古丝绸之路传到了新疆以及其他地区。

In addition, traces of rhomboid chain like Leno were also found in the No. 30 tomb in Alagou, Tianshan, Turpan Basin. The Hua Leno fabric unearthed in Minfeng, Xinjiang was also woven from this structure, with a warp density of 66 pieces/cm and a weft density of 26 pieces/cm. The diamond patterned silk fabrics unearthed from the Eastern Han tombs of Noin Ula in Mongolia and King Xu of the Eastern Han Dynasty in Korea were also the same as this silk fabric. These places were not ancient thoroughfares, but also unearthed woven fabrics from the Central Plains, indicating that the woven fabrics from this period had been transmitted to Xinjiang and other regions through the ancient Silk Road.

第五节　　隋唐时期的罗/Leno in the Sui and Tang Dynasties

隋朝的建立结束了国内分裂的局势，重新建立了统一的封建王朝，农业、手工业有了较大的发展，尤其是纺织手工业有了突出的进步，主要集中在河南、河北、四川、山东一带。到了唐朝，官营和私营的纺织业规模都相当大，罗织物的生产技术已经十分成熟，不仅生产数量巨大，而且组织变化丰富。根据《唐六典》的记载，唐朝的官营手工业规模庞大，分工精细，其中就包含有专门生产罗织物的罗作。到了唐朝后期，经济重心逐渐向南发展，浙东"机杼耕稼，提封七州，其间茧税鱼盐衣食半天下"，说明唐朝后期，江南地区丝织业的地位已经大大提升，成为全国丝织的重要区域。

The establishment of the Sui Dynasty put an end to the domestic division and reestablished a unified feudal dynasty. Agriculture and handicrafts made significant progress, especially in the textile handicraft industry, which was mainly concentrated in Henan, Hebei, Sichuan, and Shandong. In the Tang Dynasty, both state–owned and private textile industries had a considerable scale, and the production technology of woven fabrics was also very mature. Not only was the production quantity huge, but the organizational changes were more diverse. According to the records in the *Six Classics of the Tang Dynasty*, the official handicrafts of the Tang Dynasty were of a large scale and finely divided, including the production of leno fabrics. In the later period of the Tang Dynasty, the economic center gradually developed southward. Zhejiang's eastern region cultivated crops with machinery and was granted seven prefectures, during which cocoon taxes, fish, salt, clothing, and food were distributed throughout the country. It indicated that in the later period of the Tang Dynasty, the status of the silk weaving industry in the Jiangnan region had greatly improved, becoming an important region for silk weaving in the country.

这一时期最具代表性的丝绸产品为越州出产的越罗。从唐代起，江南道的越州出产的罗一跃成为全国著名的丝绸产品。历史上以地域驰名的丝织品不少，属于中原的有齐纨、鲁编、襄邑美锦，属于四川的有蜀锦，属于江南地区且被冠以越、吴之名的丝织品主要有越罗与吴绫。吴绫的生产地域较广，当时的越州、杭州、湖州均生产吴绫，江苏是吴绫的主要产地。唐朝时期，越罗、吴绫可与成都的蜀锦相提并论。

The most representative Silk product of this period was the Yue Leno produced in Yuezhou. That is to say, since the Tang Dynasty, leno fabrics produced in Yuezhou, Jiangnan Prefecture (now Shaoxing City, Zhejiang Province), have become famous silk products nationwide. In history, there were many well-known silk fabrics named after their county of origin, for example, in the Central Plains there were Qi Wan (fine silk fabrics of Qi), Lu Bian (braided fabrics of Lu), and Xiangyi

Brocade, in Sichuan Province there was Shu Brocade, and in Jiangnan region there were Yue Leno and Wu Damask Silk fabrics. The production area of Wu Damask Silk fabrics was relatively wider, including cities like Yuezhou, Hangzhou, and Huzhou, which made Jiangsu the main production area of Wu Damask Silk fabrics. In the Tang Dynasty, Yue Leno and Wu Damask Silk fabrics were comparable to Chengdu's Shu brocade.

唐代诗人杜甫在《白丝行》一诗中写道："缫须长不须白，越罗蜀锦金粟尺。"意思是说，像越罗与蜀锦这样的珍贵织物，必须用金粟尺来量取。金粟尺是唐代富贵人家的用品，尺上的星点用金粟嵌成。如果将蜀锦和越罗这两种驰名天下的产品做一比较，那么，蜀锦厚重、华丽，越罗则轻柔、曼妙；蜀锦常用作宫殿华屋的装饰物、达官贵人的袍料，越罗则是有江南韵味的丝织品，仿佛只与江南的风景与人物相配衬。

The famous poet Du Fu of the Tang Dynasty wrote in the poem *White Silk*: " Silk reeled from cocoons needs to be long but not pure in whiteness; Yue Leno and Shu Brocade use a foot-long ruler marked with golden grains," which reveals that high quality silk products like Yue Leno and Shu Brocade were so precious that only rich families could afford. If you compare the two renowned silk products, Shu Brocade and Yue Leno, you will find Shu Brocade thick and resplendent, while Yue Leno soft and elegant. Shu Brocade has been used for the decoration of palaces and the garments of dignitaries and nobles, but Yue Leno fabrics had the charm Jiangnan region, which seems to only match with the scenery and characters of Jiangnan.

杜甫在《白丝行》中咏越罗："春天衣著为君舞，蛱蝶飞来黄鹂语。落絮游丝亦有情，随风照日宜轻举。"女子穿着越罗轻舞飞扬，鸟蝶花絮也随之而舞。唐代诗人刘禹锡在《酬乐天衫酒见寄》中写道："酒法众传吴米好，舞衣偏尚越罗轻。动摇浮蚁香浓甚，装束轻鸿意态生。"刘禹锡以此来表达收到好友白居易送来的越罗和吴酒后欣喜不已的心情。越罗的轻盈更让人联想到诗人"装束轻鸿"、逸态横生的模样。诗人张泌在诗中写道："燕双飞，莺百转，越波堤下长桥。斗钿花筐金匣恰，舞衣罗薄纤腰。"李商隐、李贺等著名诗人，也都将越罗入句，描写身着越罗的少年。越罗唯美、珍贵，却不沾染人间的富贵气。罗衣、罗裳、罗裙、罗袖、罗带、罗袜、罗帐均为纤丽之物。

In *White Silk* by Dufu, the beauty wearing Yue Leno dresses dances for her lover on a day of spring, attracting flying butterflies and singing orioles to join, which embodies the fineness of Yue Leno fabrics. After receiving Yue Leno fabrics and Wu wine from his friend, poet Bai Juyi, poet Liu Yuxi described that the mellow fragrance of Wu wine and the lightness of Yue Leno. The poetry gives us the carefree image of the poet when wearing light costume. Poet Zhang Mi expressed the beauty of Yue State like this— "In the spring with birds dancing and singing, on the bridge connecting river walls, a beauty wearing exquisite jewelry, and leno dress highlights her slender waist." Li Shangyin, Li He, and other famous poets also mentioned Yue Leno in poems by describing youths dressed in Yue Leno costumes. And Yue Leno is so beautiful, it's precious but not meretricious. Those garments, skirts, sleeves, belts, socks, and curtains made of leno were all delicate and fine.

这一时期，在我国西北地区还出现了一些罗织物，这些罗织物的组织结构和秦汉时期的菱纹罗相同，均以非固定绞组的四经绞罗为地，二经绞罗为花，只是织物的纹样有所变化，为小几何花纹，织物的密度相对较大，质地较密。

During this period, some silk fabrics were also produced in the northwest region of China. The organizational structure of these silk fabrics was the same as that of the diamond patterned silk fabrics from the Qin and Han Dynasties, both of which were made of non-fixed four warp twisted silk as the ground and two warp twisted silk as the flowers. However, the pattern of the fabric changed to small geometric patterns, with a relatively high density and dense texture.

可以看出，在隋唐时期，罗织物的组织结构发生了较大的变化，除了延续早期的非固定绞组罗织物外，还出现了新的品种，即有固定绞组的罗。

It can be seen that during the Sui and Tang Dynasties, the organizational structure of leno fabrics underwent significant changes. In addition to continuing the early non-fixed twisted leno fabrics, new varieties emerged, namely leno with fixed twisted groups.

第六节　宋元时期的罗/Leno in the Song and Yuan Dynasties

宋朝时期，商品经济空前繁荣，纺织手工业也在唐朝的基础上得到进一步发展。当时丝绸的生产仍然保持和唐朝相似的形式，分为官营、私营和官雇民机包织三种。除京城之外，宋朝还在当时丝绸生产发达的地区设立生产作坊，规模巨大。北宋中期在润州（今镇江）设立了"织罗务"，专门为朝廷生产御用的罗织物，每年产量多达万余匹。民间的丝织业在当时也有"连甍比室，运箴弄杼""竹窗轧轧，寒丝手拨"的盛况。除此之外，还出现了专门从事纺织生产的家庭作坊——机户，即由官府提供原料，机户承包织造，最后产品由官府统一收购。宋朝是罗织物发展的高峰时期，不仅产量巨大，织造技术也发生了革命性的变化，有固定绞组的罗织物已经成为当时的主流。

During the Song Dynasty, the commodity economy prospered. The textile handicrafts developed further on the basis of the Tang Dynasty. At that time, silk production still maintained a similar form to that of the Tang Dynasty, which was divided into three types: official, private and civilian machine hired by official. In the Song Dynasty, in addition to the capital, large-scale workshops were set up in areas where silk production was well developed. In the middle of the Northern Song Dynasty, "Weaving Leno Service" was set up in Runzhou (now Zhenjiang) to produce the imperial leno for the court, with an annual output of more than ten thousand pieces. At that time, the folk silk weaving industry reached a prosperous situation — "the houses are close together and the workers inside are constantly weaving" "bamboo window is rolling, workers' hands are plucking the cold silk threads". In addition, there were also family workshops specializing in textile production—machine households, which were in the form of government-hired private machine weaving. The government provided raw materials, while machine households contracted to weave. The final products were purchased by the government. During the Song Dynasty, the development of leno weaving reached its peak. Not only the output of leno was huge, but also the weaving technology had a revolutionary change. Leno with fixed set had become the mainstream at that time.

相对于传统的非固定绞组的花罗，地部组织和花部组织均为链式结构，而有固定绞组罗织物的地部组织为绞状，花部组织则更多以简单的平纹组织、斜纹组织或经纬线浮长呈现，花部组织与地部组织之间的反差效果更大，从而使花部图案更加明显。又因为有固定绞组的花罗织物可在一般的花楼提花机生产，织物的幅宽内提花经线数量没有限制，因此花部可进行大面积的提花，这相对于早期的几何小花纹有很大的进步。这一时期织物的纹样主要是各种花卉图案，如牡丹、月季、芙蓉、蔷薇等，花纹图案更加精致复杂。图1-4所示是福建福州南宋黄昇墓出土的一件花罗织物。

Compared with the traditional jacquard leno in a non-fixed set, the ground structure and jacquard structure are both chain-like structures. The ground structure of the fixed leno set is twisted, and the jacquard structure is most in simple plain weave, twill weave or warp and weft float length. The contrast between the jacquard structure and ground structure is greater, thus making the jacquard pattern more obvious. Because the jacquard leno with the fixed set can be produced in the general jacquard looms and the number of jacquard warps within the width of the fabric is not limited, a large area can be jacquard, which is a great progress compared to the early geometric small pattern. The pattern of this period is mainly various floral patterns, such as peony, Chinese rose, hibiscus and rose. The pattern is more exquisite and complex. Figure 1-4 shows a jacquard leno excavated from Huang Sheng's tomb of the Southern Song Dynasty in Fuzhou City, Fujian Province.

图1-4　宋代花罗织物纹样
Jacquard Leno Pattern of the Song Dynasty

元朝是蒙古游牧民族建立的王朝，马背上的英雄更爱织金锦的豪华，不太能理解罗之美。由于对丝织业赋税太重，加上水利失修等种种原因，元代的丝绸业经历了艰难的发展。元代丝绸的产区主要集中在以腹里区为中心的黄河下游和以浙江省为中心的长江下游地区，而其他地区的丝绸业已经逐渐衰落。

The Yuan Dynasty was established by Mongolian nomads. The heroes on horseback were less appreciative of leno, but more in love with the luxurious Nasich. Due to the heavy tax on the silk industry, water conservancy disrepair and other causes, the silk industry in the Yuan Dynasty experienced difficult development. The silk production areas of the Yuan Dynasty were mainly concentrated in the lower reaches of the Yellow River centered around the belly region and the lower reaches of the Yangtze River centered around Zhejiang Province, while the silk industry in other regions had gradually declined.

但是罗织物仍然是元代统治者大量使用的织物。在《元典章》中记载，当时文武百官的官服都用罗织物制成，王室贵族的幕帐等生活用品也多用罗织物。

However, woven fabrics were still widely used by the rulers of the Yuan Dynasty. In the *Yuan Dianzhang*, it is recorded that at that time, the official uniforms of civil and military officials were all made of woven fabrics, and the daily necessities of the royal nobility, such as curtains and tents, were also mostly made of woven fabrics.

此外，由于元代统治者喜金，织金、印金等工艺也常常使用在罗织物上。织金是将金打成金箔后捻成金线，或者将金箔直接切成金丝，作为经纬线织入罗织物中，从而使织物的表面色泽更加艳丽。印金是以素罗织物作为底料，在上面用凸版印上粘合剂后再撒上金粉，印金可以使用在整件衣服上面。宋代的时候虽然也出现了印金的罗织物，但是都只是用于衣服的局部位置。这些织金、印金的罗织物都是以添加不同成分的材料来提升罗织物的价值和档次，比普通罗织物更加名贵。

In addition, due to the rulers of the Yuan Dynasty's preference for gold, techniques such as weaving and printing gold were often used on woven fabrics. Weaving gold is the process of beating gold into gold foil, twisting it into gold thread, or directly cutting the gold foil into gold thread, weaving it into woven fabric as warp and weft threads, thereby making the surface of the fabric look more colorful. Printing gold is a technique that uses plain silk fabric as the base material, with adhesive printed on it using relief printing and then sprinkled with gold powder. Printing gold can be used on the entire piece of clothing. During the Song Dynasty, although gold printed fabrics also appeared, they were only used for local positions in clothing. These woven and printed gold fabrics are made by adding materials with different compositions to enhance their value and grade, making them more precious than ordinary fabrics.

第七节　明清时期的罗/Leno in the Ming and Qing Dynasties

明清时期是我国的手工织造技术发展的巅峰时期，统治阶级在全国范围内广泛开展蚕桑丝织业，许多盛产丝绸的地区都成了当时兴旺发达的城市。江南地区从宋朝开始就已经成为丝绸生产的重心，如江苏的苏州、南京、无锡，浙江的杭州、湖州、嘉兴等城镇，可谓桑麻遍地，户户机声，丝绸产品无论是质量还是产量都位居全国之首，蚕桑丝绸的生产已经成为江南地区重要的经济收入和生产行业。

The Ming and Qing Dynasties were the peak period for the development of handmade weaving technology in China. The ruling class extensively engaged in sericulture and silk weaving throughout the country, and many areas rich in silk became prosperous and developed cities at that time. Since the Song Dynasty, the Jiangnan region had been the center of silk production, such as Suzhou, Nanjing, Wuxi in Jiangsu, Hangzhou, Huzhou, Jiaxing in Zhejiang, and other cities. It can be said that mulberry and hemp are everywhere, and every household produces silk products with the highest quality and output in the country. The production of sericulture and silk has become an important economic income and production industry in the Jiangnan region.

明清时期的统治阶级喜好华丽奢侈的风格，对丝绸产品的图案、色彩、质量等都有更高的要求。丝绸生产者也不断提高织造的技术，增加花色的品种，提高丝绸产品的质量。

The ruling class of the Ming and Qing Dynasties favored luxurious style, and had higher requirements for the patterns, colors, and quality of silk products. Silk producers were also constantly improving their weaving techniques, increasing the variety of patterns, and improving the quality of silk products.

就罗织物而言，当时主要分为三梭罗、五梭罗、素罗、花罗等品种，而传统的链式罗已经非常少见。这主要是建立在社会对罗织物需求量大大提高的基础之上，人们为了快速、大量地生产罗织物，不得不放弃生产效率较低的链式罗，从而导致链式罗逐渐消失，而固定绞组罗织物的发展成了主流。

As far as leno fabric was concerned, it was mainly divided into three types of leno, five types of leno, Su Leno, Hua Leno, and other varieties, while traditional chain leno was already very rare. This was mainly based on the significant increase in demand for woven fabrics in society. In order to produce woven fabrics quickly and in large quantities, people had to give up the chain type woven fabric with lower production efficiency, which gradually led to the disappearance of chain type woven fabric, and the development of fixed twisted group woven fabric became mainstream.

当时生产的罗以杭州地区的最佳，称为杭罗。杭罗是不同于越罗的一种新型罗织物，然而，在先秦至元代的历史长河中，链式罗一直都占有罗织物主流的地位，固定绞组的杭罗却很少。杭罗作为固定绞组罗的延续，其技术基础可追溯至新石器时代晚期。然而，杭罗经过数千年的沉寂后，随着唐宋时期江南地区开发的深入而迎来发展的新时期，并且固定绞组的杭罗在明代逐渐发展成为罗织物的主要组织形式，完全取代了链式罗。从纺织技术的角度看，杭罗的生产是一种技术上的退化，即简单的织造技术取代复杂的织造技术。从社会因素的角度看，一方面，江南地区商品经济的发展为杭罗的发展提供了强大的驱动力；另一方面，杭州城市的发展最终使杭罗完成技术上的集中，使其成为具有杭州地方特色的一种丝织物。

At that time, the best leno produced in the Hangzhou area was called Hang Leno. Hang Leno was a new type of leno fabric different from the Yue Leno. However, in the long history from the pre-Qin Dynasty to the Yuan Dynasty, the chain-like leno had always occupied the mainstream of leno fabric. Leno fabric with structure was seldom. Hang Leno, as a continuation of fixed twisted set leno, has a technical basis dating back to the late Neolithic period. After thousands of years of silence, it ushered in a new period of development with the intensive development of Jiangnan region during the Tang and Song Dynasties. During the Ming Dynasty, it gradually became the main form of the leno fabric, completely replacing the chain-like leno. From the perspective of textile technology, the emergence of Hang Leno is a kind of technical degradation, that is, simple weaving technology replaces the complex. From the perspective of the social factors, on the one hand, the development of the commodity economy in Jiangnan region provided a strong driving force for the development of Hang Leno; on the other hand, the development of Hangzhou eventually made Hang Leno complete the technical concentration, making it become a kind of silk fabric with local characteristics of Hangzhou.

明代时期，杭州是浙江布政使司的驻地，经济非常繁荣，明末丝织行业就产生了资本主义萌芽，商品经济高度发达又使当地的社会风气更加世俗化和追求享乐，皇室、官僚享用的罗织物开始向民间扩散。价格更便宜、结构更牢固的固定绞组罗焕发出新生，拥有了更为广阔的消费市场。

In the Ming Dynasty, Hangzhou was the residence of the Provincial Administrative Government, with prosperous economy. In the late Ming Dynasty, the origin of capitalism was born in the silk weaving industry. Likewise, the highly developed commodity economy made the local social atmosphere more secular and pleasure-seeking. As a result, the leno fabrics, which used to be enjoyed only by the royal family and bureaucrats began to spread to common people. The cheaper and more solidly constructed fixed set leno began to enjoy a broader consumer market.

此外，明清时期的刺绣、织金、妆花、印染等精细装饰加工工艺也都发展到了极高的水平。因此，罗织物也被作为基础面料进行装饰，织金罗、刺绣罗、妆花罗均是当时名贵的面料，深受皇室阶级喜爱。

In addition, fine decorative processing techniques such as embroidery, gold weaving, flower making, and printing and dyeing during the Ming and Qing Dynasties also developed to a very high level. Therefore, leno fabric was also used as a basic fabric for decoration. Gold Weaving Leno, Embroidery Leno, and Makeup Flower Leno were all precious fabrics loved by the royal class at that time.

○ 第二章

罗的织造技艺
The Weaving Technology of Leno

织造技艺是织物结构的实现方式，历史上绞经织物在组织结构、花型纹样、织物风格上的发展，其背后都是织造技艺的进展。绞经织物的织造流程与其他丝织物基本一致，但相对于普通的丝绸产品，罗织物由于其结构的特殊性，织造技艺也更为复杂。

Weaving technology is the way to the actual realization of fabric structure. Historically, the development of strand textile in terms of the weave structures, pattern designs and fabric styles attributes to the upgrading of weaving technology. The weaving process of strand textile is basically the same as that of other silk fabrics, but compared with ordinary silk products, the weaving technology of leno is more complicated due to its unique structure.

第一节　罗织造技艺的演变/The Evolution of Leno Weaving Technology

从纺织史的研究角度来看，纺织品的起源时常和编织物相联系，机械织造的平纹织物结构来源于手工编织物结构。绞经织物的组织结构与手工编织物的结构也很相近，史前墓葬中出土的几块罗织物的结构也介于手工编织物结构和绞经织物结构之间。

Throughout the textile history, the origin of textiles was derived from braided fabrics, and woven plain cloth from hand knitting fabrics. The structure of strand textile is obviously more consistent with that of braided fabrics. Moreover, the structure of several leno fabrics excavated in prehistoric tombs also resembles the hand knitting fabrics and strand textile.

一、史前的绞经与绞编织造技艺/The Twist Knitting Weaving Technology of Strand Textiles in Prehistory

根据考古发现，我国自新石器时期就存在绞编织物。我国出土的最早的绞编织物是江苏吴县草鞋山新石器时期遗址（距今约6000年）中的三块野生葛织成的罗织物，如图2-1所示。这几块织物为纬起花的罗纹织物，织品已经炭化，织物组织结构为绞纱罗纹，嵌入绕环斜纹,还有罗纹边组织。织物伸直的经纱相互平行,纬纱按照二上二下的规律缠绕,与经纱进行交织,每个绞组包含两根纬纱,且同一绞组的两根纬纱相互扭绞。在显花部分,每根纬线在经线上缠绕一圈,构成菱形和山形的纹路。残片的一边有山形和菱形斜纹,花纹处的纬纱曲折变化,罗纹部纬纱是上下绞结。经纱为双股,S捻。经密约10根/cm,罗纹部纬密26～28根/cm,地部纬密13～14根/cm。根据其起花处不规则的绞缠方式,可以判断这块织物属于手工绞编的原始织物。

According to archaeological findings, strand textile has existed in China since the Neolithic Age. The earliest discovery was three pieces of strand textiles made by wild Kudzu in the Neolithic Site of Caoxie Mountain (about 6,000 years ago) in Wuxian County, Jiangsu Province, as shown in Figure 2-1. These carbonized pieces were weft-knitted rib fabric, whose structures were skein patterns. They were embedded with twill weave, having rib stitch in the edge. The warp yarns were parallel to each other, interleaved with the weft which were wrapped around within the rule of two up and two down. Each skein group contained two weft yarns, twisting with each other. For the patterns, each weft yarn was wrapped in a circle on the warp to form diamond and mountain shapes. On one side of the fragment, a mountain and diamond patterned twill weave could be found. The weft yarn at the pattern changed in twists and turns, and stranded up and down at the rib stitch. The warp was double stranded and S-twisted. The warp density was about 10 pieces/cm, the weft density of the pattern part was 26-28 pieces/cm, and the weft density of the ground part was 13-14 pieces/cm. By the irregular strand method of the pattern, we could identify this fabric to be the original hand-knitted strand textile.

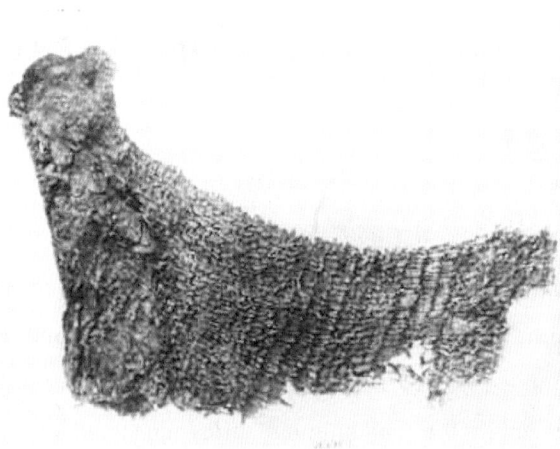

图2-1 江苏吴县草鞋山出土的葛制的绞罗织物
Strand Leno Textiles Made by Kudzu, Excavated in Caoxie
Mountain in Wuxian County, Jiangsu Province

　　紧随其后,西安半坡仰韶文化遗址(距今约6000年)中也发现了绞经织物存在的痕迹。遗址中虽然没有直接出土绞经织物实物,但出土了大量的陶器,在陶器表面存在很多织物印痕,其中大多为平纹织物印痕,少数为绞经织物印痕。考古发现的早期绞经织物经历了几千年的历史，有的炭化，有的只有残存的痕迹，借助更易保存的物品才得以追寻到一丝踪迹，虽然不是直接的织物样本，但对古代纺织品的研究来说仍然是宝贵的资料。河南荥阳青台遗址（距今约5500年）中发现了属于仰韶文化时期的绞经织物，如图2-2所示。织物为浅绛色，以蚕丝为原料，织物组织为二经绞罗纹，采用左右互相绞转的方式编织。青台遗址中还发掘了纺轮、骨针、陶匕等与纺织相关的遗存，可能与当时绞经织物的织造有关。

Subsequently, strand textiles were also discovered in the Banpo Site of Yangshao Culture in Xi'an（about 6,000 years ago）. Although there were no existed strand textiles in the ruins, a large number of potteries were excavated, and there were many fabric imprints on the surface of these potteries. Most of them were plain woven fabrics, and a small amount were strand textiles. With a thousand years of history, some of the primitive strand textiles discovered by archaeology were carbonized, and some were only left in pieces and can only be traced with the help of more easily preserved media. Although they were not actual strand textiles samples, they were still invaluable for the research. In the Qingtai Relic of Xingyang City, Henan Province（about 5,500 years ago）, strand textiles belonging to the Yangshao Culture were discovered, as shown in Figure 2-2. The fabric was light crimson, with silk as the material. Its structure was a twisted rib, whose left and right warp coiled around. Other textile-related relics were also excavated in the site, such as the spinning wheel, bone needle and pottery dagger, which may be connected with strand textiles of the time.

图2-2　河南荥阳青台遗址出土的绞经织物
Strand Textiles Excavated in Qingtai Relic of Xingyang
City, Henan Province

二、商周时期的绞经织造技艺/The Weaving Technology of Strand Textiles in the Shang and Zhou Dynasties

殷商时期著名的殷墟妇好墓中同样发现了绞经织物的踪迹。墓中出土了绞经组织的大孔罗织物，织物形成的孔眼较大。墓葬出土的大孔罗织物有两件，保存较好的是在一件妇好连体甗的口下，织物痕迹分布面积较大；另一件保存在铜小方彝的盖上，如图 2-3 所示。大孔罗织物的经纱宽度为 0.12～0.15mm，纬纱宽度为 0.12mm，经纱密度为 32 根 /cm，纬纱密度为 12 根 /cm，经纬纱均为 S 捻，约捻度为 1512 捻 /m。

Strand textiles were also discovered in the tomb of Fu Hao Yin Ruins, a famous tomb in the Shang Dynasty. The holes formed by those fabrics were large. There were two pieces of leno fabrics with large holes excavated from the tomb. The better preserved was under the bottle of the Yan, a bronze conjoined ancient kitchen utensil, where large fabric traces were left. The other piece was kept on the cover of the Yi, a copper square wine container, as shown in Figure 2-3. The warp width was 0.12-0.15 mm, and the weft width was 0.12 mm. The warp density was 32 pieces/cm, the weft density was 12 pieces/cm. Both warp and weft yarns tended to be S-twisted, about 1,512 twist/m.

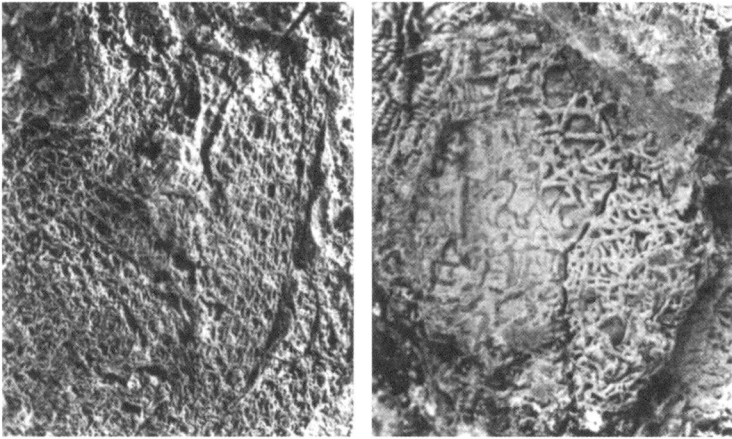

图 2-3　河南殷墟妇好墓中青铜器表面的罗织物遗迹
Leno Remains on the Surface of Bronze Container, Excavated in Fu Hao Yin Ruins, Henan Province

进入周代，绞经织物的种类开始增多，但大多为织物结构简单的绞经织物。例如，在河南省三门峡市上村岭虢国国君虢仲墓中出土了衣服残片，其中残留了一些织物标本，在棺外，有两件套穿在一起的合裆麻裤和一片矩形领口的麻上衣残片，这两件织物保存相对完整。在棺内，有部分衣服残片，随葬玉饰穿系中还残留了一些织物标本，包含了绮、绢、组、绣、罗、印绘等织物种类。

In the Zhou Dynasty, the types of strand textile began to increase, but there was still a large proportion of simple strand textile. For example, some specimens remained in the clothes excavated in the tomb of Guo Zhong, king of the Guo State in Shangcunling, Sanmenxia City, Henan Province. Outside the coffin, there were two stacked flax pants and a piece of flax jacket with rectangular neckline. These two fabrics were relatively intact. Inside the coffin, there were some clothes fragments, and some fabric specimens were in the burial jade ornaments, including figured woven silk material, waste silk, set-up, embroidery, leno, printing and painting, etc.

商周时期的织造技艺和织机已经不断趋于完善，在之前原始腰机的技术上有很大的发展。从史前到周代早期，绞经织物的结构逐渐稳定下来，以简单的二经绞织物为主，织造技艺从以手工为主发展到以简单织机为主的固定模式，织物随着织造技艺的发展逐渐变得精细。

The weaving technology and looms in the Shang and Zhou Dynasties had been constantly improved, which had a great development in the original waist machine. From prehistory to the early Zhou Dynasty, the structure of strand textiles gradually stabilized, dominated by twisted ribs. The upgrading of weaving technology which developed from hand-knitting to simple looms gradually resulted in more exquisite fabrics.

三、战国至秦汉时期的绞经织造技艺/The Weaving Technology of Strand Textiles from the Warring States Period to the Qin and Han Dynasties

战国至秦汉时期，丝绸织造技艺显著提高，织造机器不断创新，丝绸产品产量大幅提高。

During the Warring States Period and the Qin and Han Dynasties, silk industry made remarkable progress in the improvement of weaving technology, the innovation of weaving looms, and the significant increase in production quantity of silk fabrics.

战国时期，当时主要的织机类型为脚踏提综开口的斜织机，它将提综动作从手提改进为脚踏，织工的手可以从事投纬、打纬，提高了生产效率。从汉代保存的画像石上能够清楚地看到当时斜织机的样式，如图2-4所示。据统计，我国先后出土了主要描绘当时普通家庭纺织生产的汉代画像石约18块。

During the Warring States Period, the petal twill tape loom was the mainstream, which improved the shedding process from hand lift to foot lift. The weaver's hands could be engaged in picking and beating, largely increasing productivity. On the stone relief of the Han Dynasty, the twill tape loom can be clearly identified, as shown in Figure 2-4. According to statistics, China has unearthed about 18 stone reliefs of the Han Dynasty, mainly depicting the textile production of ordinary households at that time.

虽然汉代画像石中都是简单的织机，但是由于这一时期已经出现了大量的提花织物，如湖南长沙马王堆一号汉墓出土的菱纹罗，可以看出提花织机及提花织造技艺在这一时期已经出现，主要是由原始的提花腰机发展形成的多综多蹑织机。《西京杂记》中也提到当时有一百二十蹑织机，这证明汉代时综蹑提花织机已经得到普及。以马王堆出土的菱纹罗为例，其地部为四经绞罗，花部为二经绞罗。织造这样的组织时，需要在织机上装造地综、绞综和纹综，地综和绞综分别为两片，纹综根据织物循

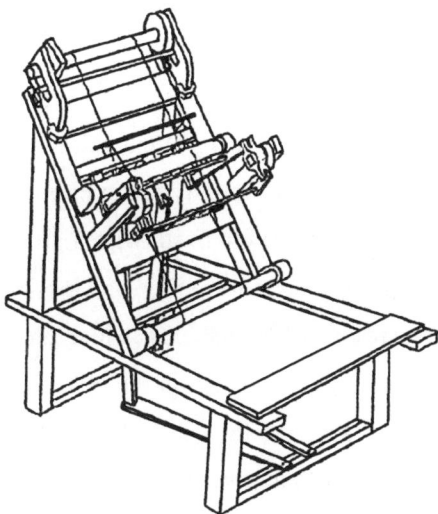

图2-4　斜织机
Twill Tape Loom

环内纬线数，并结合图案对称的特点，选择54片纹综。

The looms on the stone reliefs were simple types. However, a large number of jacquard fabrics had been excavated in this period, such as the diamond leno excavated from the Tomb No. 1 at Mawangdui in Changsha City, Hunan Province. Therefore, we can conclude that jacquard looms and weaving technology had been invented at that time. The loom with multiple healds and crepes mainly developed from the original jacquard waist machine. As mentioned in *Miscellanea of the Western Capital*, there were 120 crepe looms at that time, which proved that the jacquard loom had been popularized in the Han Dynasty. Taking the diamond leno of Mawangdui as an example, it had four warp twisted ribs on the ground part and two warp twisted ribs on the pattern. To weave such fabric, it was necessary to install 2 pieces of ground healds, 2 pieces of twisted healds and 54 pattern cards on the loom. The pattern cards were selected by the number of weft yards and features of pattern symmetry.

四、隋唐时期的罗织造技艺/The Weaving Technology of Leno Fabrics in the Sui and Tang Dynasties

隋唐时期，罗织物的组织结构发生了较大的变化，除了延续早期的非固定绞组罗织物外，还出现了新的品种，即有固定绞组的罗。

In the Sui and Tang Dynasties, the weave structure of leno changed greatly. In addition to the previous non-fixed leno fabrics, new varieties that were the fixed leno fabrics also had been produced.

敦煌莫高窟中发现了一块晚唐至五代时期的紫色联珠方格"卍"字纹暗花纱，如图2-5所示，属于有固定绞组的二经绞平纹提花织物，地部由绞经和地经对称绞织而成，花部是不起绞的平纹组织。织物纹样由联珠方格和"卍"字纹组成，现收藏于大英博物馆中。我国西北地区还出土了一些隋唐时期的罗织物——织花罗，这些罗织物的组织结构和秦汉时期的菱纹罗相同，均以非固定绞组的四经绞罗为地，二经绞罗为花，只是织物的纹样有所变化，为小几何花纹，如图2-6所示。

In the Dunhuang Mogao Grottoes, a piece of purple beaded shadow yarn with "卍" pattern in the late Tang Dynasty to the Five Dynasties was found, as shown in Figure 2-5. It was the plain jacquard fabric with fixed twisted groups. The ground part was symmetrically formed by the strand warp and ground warp. The pattern was made by plain weave. The fabric was composed of beaded squares and "卍" patterns, which are now collected in the British Museum. There are also other leno fabrics excavated in the Northwest China, whose weave structures were the same as that of diamond leno in the Qin and Han Dynasties. They all had non-fixed four warp twisted ribs on the ground part and two warp twisted ribs on the pattern. The patterns varied differently with small geometric figures, as shown in Figure 2-6.

图 2-5 紫色联珠方格 "卍" 字纹暗花纱
Purple Beaded Shadow Yarn with "卍" Pattern

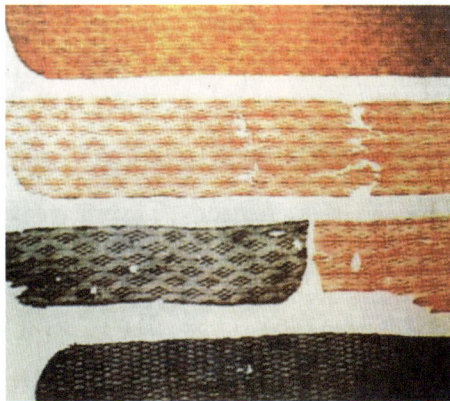

图 2-6 织花罗
Jacquard Leno Patterns

五、元时期的罗织造技艺/The Weaving Technology of Leno Fabrics in the Yuan Dynasty

通过文献记载，元代丝绸的产区主要集中在以腹里地区为中心的黄河下游和以浙江省为中心的长江下游地区，而其他地区的丝绸业已经逐渐衰落。从组织结构来看，传统的非固定绞组罗织物仍占有一定的地位，有固定绞组的罗织物同期大量的盛行。除了宋代的提花罗，元代还出现了新的横罗织物，即在绞经和地经绞转之后，织入一定奇数梭的平纹组织，从而在横向上形成条状的空隙，根据织入的纬数可以分为三梭罗、五梭罗等。

According to literature records, the silk production areas in the Yuan Dynasty were mainly concentrated in the lower reaches of the Yellow River centered on the Central region of the Yuan Dynasty and the lower reaches of the Yangtze River centered on Zhejiang Province. However, the silk industry in other areas had gradually declined. From the view of the weave structure, the traditional non–fixed leno fabric still had the upper hand in the market, while the fixed leno fabric obviously prevailed. In addition to the jacquard fabric in the Song Dynasty, a new horizontal leno fabric that was the plain stitch appeared in the Yuan Dynasty. After twisting the strand warp and ground warp, a certain odd shuttle was woven so as to form a strip gap in the horizontal direction. By the number of weft, it could be divided into three shuttle leno, five shuttle leno and so on.

这一时期还出现了一种特殊的绉罗（图2-7），即在绞转后织入偶数梭平纹，不共口的结构使梭口处的经线滑移，织物表面形成绉效应。如山东邹县出土的鱼莲绉罗，在梭口处绞经、地经和纬线重叠，织物表面凹凸不平，与绉织物相似，这种织物组织是非常少见的。

图 2-7 元代的绉罗示意图
Crepe Leno in the Yuan Dynasty

23

During this period, a special crepe leno（Figure 2-7）was produced. A certain even shuttle plain stitch was woven after twisting. The structure of different openings made the warp slip at the shed, forming the crepe effect on the surface. For example, the crepe leno with fish and lotus patterns unearthed in Zouxian County, Shandong Province had the similar structure. The strand warp, ground warp and weft overlapped at the shed, and the surface of the fabric was uneven. This type of weave structure is very rare.

六、明清时期的罗织造技艺/The Weaving Technology of Leno Fabrics in the Ming and Qing Dynasties

明清时期的织造机械已经发展得很完善，并且针对不同的织物有不同的织机类型。明代宋应星在《天工开物》中写道："凡罗，中空小路，以透风凉，其消息全在软综之中。衮头两扇打综，一软一硬，凡五梭三梭之后，踏起软综，自然纠转诸经，空路不粘。"这段文字描述的就是织造三梭罗、五梭罗的过程。从文字中可以看出，罗织物起绞的关键就是软综，即绞综。因此，在普通的织机上加上绞综装置，使经线在提升时带动绞经扭转，便能够织造罗织物。再结合素综织造平纹组织，则可织造出具有独特效果的杭罗织物。除了素罗，提花杭罗在明清时期也是非常重要的品种，它就是在杭罗织物的平纹地上通过越纬或者起斜纹小花来形成图案，因此只需在提花织机上"加桄二扇"，也就是加装提花楼提花即可。明清时期的杭罗织机与两宋时期的花罗织机在外形和结构上基本保持一致，是在之前织机的基础上，再结合对应的织造工艺发展形成的。

The weaving looms in the Ming and Qing Dynasties had been well developed, and divided into different types for different fabrics. Song Yingxing in the Ming Dynasty wrote in *Exploitation of the Works of Nature* that "Leno has so many holes that is breathable. The key lies in the soft heald, or the twisted heald. The two healds, one soft and one hard were used in the process. After three or five shuttles, the soft heald was released to drive the strand warp." It describes the process of weaving three-shuttle leno and five-shuttle leno. It can be seen that the key to weave the leno fabric is the soft heald. Therefore, as long as the twisted heald device is installed on the loom, the warp is able to drive the strand fabric when lifted, and then the leno fabric can be made. Combined with plain weave, Hang Leno fabric with unique effect can be produced. Besides plain leno, jacquard Hang Leno was also a very important type in the Ming and Qing Dynasties. Based on the plain weave in the horizontal leno fabric, the patterns were made by crossing weft or raising twill figures. Therefore, it only needs to add jacquard patterns. Hang Leno loom in the Ming and Qing Dynasties is similar to the jacquard loom in the Song Dynasty in shape and structure. It was invented on the basis of previous looms and combination with the upgrading of weaving technology.

第二节　罗生产织造的工序 /The Manufacturing and Weaving Process of Leno

根据罗织物的工艺设计，并结合苏州市锦达丝绸有限公司、杭州福兴丝绸有限公司实际生产的情况，可以将罗生产织造的工序分为原料准备、织造准备、上机织造和后整理四个部分。

According to the process design of leno and the actual production of Suzhou Jinda Silk Co. Ltd. and Hangzhou Fuxing Silk Co. Ltd., the weaving process of leno can be divided into four steps: namely, raw material preparation, weaving preparation, weaving process and post finishing.

一、原料准备 /Raw Material Preparation

原料准备是指对蚕丝进行选择和处理，使其能够满足后期罗的织造要求。原料一般选用上等蚕丝，处理前需检验蚕丝的均匀度、强度等。罗为经纬线纵横相织，但经纬之间又存在差别，一般选择最好的蚕丝作为经线，稍次一点的蚕丝作为纬线。

Raw material preparation refers to the selection and treatment of silk, so that it can meet the needs of later leno weaving. The best cocoon silk has to be selected and its uniformity and intensity shall be checked before treatment. Leno is woven by the cross of warp yarns and weft yarns, which are different from each other. Therefore, the finest silk is generally selected to do the warp and the quality of slightly worse to do the weft.

原料选择好后还需对其进行浸泡，这是因为包覆于丝素表面的丝胶会影响丝线的加工性能。浸泡的目的就是对丝线进行脱胶处理，使部分胶着的丝条松散。浸泡后的丝线柔软、光滑、利于后续的加工。其方法是将绞装的丝线浸入清水中，并加入适当助剂溶液，煮沸20min，去除纤维上的油脂和杂质，再将丝线放入清水中浸泡。

After being selected, the raw material needs to be soaked because the sericin coated on the surface will affect the processing performance of the silk. The purpose of soaking is to degum the material and loosen its strand silk. After being soaked, the silk yarn will become soft and smooth, which is conducive to subsequent processing. The specific method is to immerse the strand silk in water, and then add certain amount of auxiliaries and boil it for 20 minutes, removing the grease and impurities on the fiber, and finally, immerse the processed silk in water again.

脱胶完成后，还需对丝线进行晾晒，使其恢复平直状态。晾晒时选择通风较好的室内阴干，依靠室内温度和空气流动来使丝线中的水分挥发，使丝线自然干燥，达到合适的回潮率。除去水分之后，因蚕丝极细，绞状的蚕丝在晾干之后容易打结合拢，所以在晾干之后，需用手将蚕丝拉伸、分离，使之恢复松软状态。

After being degummed, the silk needs to be dried to restore its straightness. A well-ventilated room should be chosen to volatilize the moisture in the yarn by temperature and airflow, so that it can be naturally dried. Lastly, proper moisture regain of silk can be achieved. After volatilized and dried, the strand silk will be easy to knot because of its thinness. Therefore, it is necessary to stretch it by hand and make it soft after drying.

二、织造准备/Weaving Preparation

织造准备，即经纬线准备，是指对经过原料准备流程的丝线进行上机前的操作，主要包括络丝、并丝、整经、摇纡、穿综、穿筘和打蜡等。

Weaving preparation, namely warp and weft preparation, refers to the operation before loading on the silk thread after raw materials preparation, which mainly includes spooling, folding, warping, reeling, healding, denting and waxing, etc.

（1）络丝工艺。络丝是指将绞装的丝线卷绕到筒子或篗子上的过程，并在此过程中除去丝线上的疵点，如图2-8所示。其方法是将晾干的丝线套到翻丝车的套丝架上，然后将丝线卷绕在筒子或篗子上，通过套丝架与筒子或篗子的同时转动，将绞装的丝线络到筒子或篗子上。

图2-8　络丝
Spooling Process

Spooling technique. Spooling refers to the process of winding the strand silk onto the bobbin or swift, and removing the defects, as shown in Figure 2-8. The specific method is to put the dried silk yarn on the threading frame of the turning machine, and then wind the yarn on the bobbin or swift. By the simultaneous rotation of the threading frame and the bobbin or swift, the strand silk can be spooled on the swift.

（2）并丝工艺。并丝工艺能够使经线和纬线细度符合织造要求，使其条干更加均匀，同时还能够在此过程中清除丝线上的强力弱环，提高丝线强度。例如，杭罗的经线为3根50/70旦农工丝合成一股的纱线，纬线为4根50/70旦农工丝合成一股的纱线，因此需要并丝工艺。

Folding technique. Folding can make the yarn meet the requirements of weaving and make its evenness more uniform. Meanwhile, it can also remove the defects of strength and improve the strength of the yarn. For example, the warp of Hang Leno is composed of three 50/70D industrial filaments, and four for the weft. Therefore, the folding process is needed.

（3）整经工艺。整经是指将卷绕在筒子或篗子上的经线，按照织物规格所需要的总经丝数、门幅、长度平行地卷绕形成经轴，供后期织造使用。《天工开物》中记载着传统的分条整经的方法，现代罗织物通常采用大圆框分条整经机，即将全幅织物所需要的经线分成几条，每一条的经线密度都相同，再按照经线的长度逐条地卷绕在圆框上，然后将圆框上面的经丝卷绕成经轴，如图2-9所示。

Warping technique. Warping refers to the winding on the bobbin or swift in parallel, according to the fabric specification required for the total number of warp, width, length of warp yarns winding formation in parallel, for the use of the late weaving. *Exploitation of the Works of Nature* records the traditional method of sectional warping. Modern technology for producing leno usually adopts a large round sectional warper to divide the warp as needed to ensure each one remains the same density, then wind it on the frame one by one according to the length. Finally, a warp beam will be completed, as shown in Figure 2-9.

图2-9　分条整经机
Sectional Warper

（4）摇纡工艺。摇纡是将筒装或者籰装的丝线，卷绕成适合织造的且可以放入梭子中的纡子。纡子的质量直接影响面料的质量，所以在摇纡过程中要求张力均匀，并去除丝线中的弱节。杭州福兴丝绸有限公司保存有传统的"水纡"工艺，即将要摇纡的籰子在热水中浸泡8~9h，使之在湿态下进行摇纡，形成的纡子还要继续泡在冷水中。目的是减少织造时纡子的振动，从而使织物紧密，织纹清晰，面料表面平挺，保持良好的手感。

Reeling technique. Reeling is to wind the yarn on the bobbin or swift into a cop that suits for the shuttle. The quality of the cop will directly affect that of the fabric, so it is required to have uniform tension and remove the defects of strength during the process. Hangzhou Fuxing Silk Co. Ltd. has preserved the traditional reeling technique using water. Specifically, immerse the swift in hot water for 8-9 hours and reel them in a wet state. The formed cop will continue to be soaked in cold water. The purpose is to reduce vibration during weaving, make the fabric compact and the weave pattern clear. The fabric will be flat and straight and maintain comfortable.

（5）穿综工艺和穿筘工艺。为了满足织造需求，经线还需要穿过综框和筘齿。在一个组织循环内，提升规律相同的经线可穿入同一综片内，不同提升规律的经线需要穿入不同的综片内，如图2-10所示。通过不同综片的提升，使某一部分的经线提升，形成组织。而除了普通的综片外，杭罗的绞经还需要穿入绞综内，使经线在织造中起绞，形成杭罗独特的结构。穿完综的经线再穿入筘齿中，以满足打纬需求，使纬线更加紧密。

图2-10　整理线综
Organizing the Healds

Healding and denting techniques. To meet the weaving needs, the warp has to pass through the heald frame and reed dent. Within a round, the warp with the same lifting pace can penetrate into the same heald, while different healds for the one with different pace, as shown in Figure 2-10. By lifting different healds, the warp of a certain part can be propelled to weave. Except for the ordinary healds, the strand warp of Hang Leno also has to pierce through the twisted heald to form its unique structure. The processed warp will be threaded into the reed dent again to meet the beating demand and make the weft more compact.

（6）打蜡工艺。打蜡是指将蜡均匀地涂在经轴和经面上，起到润滑剂的作用，减少经线在织造过程中与其他部分的摩擦，减少经线断头的概率。

Waxing technique. Waxing is to apply the wax evenly on the warp beam and the warp surface, acting as a lubricant. It will reduce the friction between the warp and other parts in weaving, and reduce the chance of warp breakage.

三、上机织造 /Weaving Process

将经过准备工艺的经线安装在机器上，将经线从经轴中引出，按照一定的顺序通过绞杆，穿过综框和筘齿，经过胸梁卷绕到卷布轴上。然后将摇纡完成的纡子装入梭箱，就可以进行织造。通过脚踩机器的踏板使综框提升，从而使穿入综片对应的经线提升，形成开口。将卷绕纬线的纡管置于梭子中，梭子带领纬线穿过梭口，完成引纬。之后通过筘打纬，将引入梭口的纬线推向织口，与经线交织。通过不断的重复，纬线穿入不同的开口，与经线交织形成织物。为了保持织造的连续性，形成的织物通过卷布轴的卷取引离织口，并通过一定的张力使织物具有一定的紧密度，同时还需要通过经轴的转动，不断地放出经线来进行补充，如图2-11所示。

The processed warp will be placed on the loom and drawn it from the warp beam. It will also be passed through the crossing rod, the heald frame and reed dent in a certain order. Next, it is put on the cloth roller through the front beam. After that, the reeled cop is put into the shuttle box to carry out weaving. The loom pedal will lift the heald frame to drive the corresponding warp and form an opening. Put the cop for winding in the shuttle, which leads the weft through the shed to complete weft insertion. Then the cop will be used for weft beating and pushing the weft to the cloth fell, which will be interleaved with the warp. By constant repetition, the weft will penetrate into different openings and interweaves with the warp to form the fabric. To maintain the continuity, the formed fabric will be led away from the cloth fell through the cloth roller. The fabric will be tight through a certain tension. Meanwhile, it also needs to be supplemented by continuously releasing the warp through the rotation of the beam, as shown in Figure 2-11.

如果要在罗织物上面提花，还需要在织机上方安装提花龙头。此外，在织造过程中为了保证织物具有一定的幅宽，还需要进行"撑幅工艺"，即通过不同材质的工具在靠近织口的地方撑紧织物。

图2-11　上机织造
Weaving Process

In order to weave jacquard leno, the jacquard mechanism needs to be installed on the loom. Furthermore, in order to reach a certain width, the fabric has to be stretched. Different materials are used to help brace the fabric.

四、后整理 /Post-finishing

织造形成的杭罗织物还是粗坯，手感较硬，不能够直接使用，还需要经过进一步的脱胶和除杂处理。经过后整理的杭罗，手感柔软舒适。再按照不同的需求进行染色、漂洗、熨烫处理，从而形成精美的杭罗面料。现在生产的杭罗大部分都是素罗，因此，通常在成品的素罗面料上进行装饰工艺，如印花、染缬等，增加面料的美感。

The woven Hang Leno fabric is still coarse billet, feels hard, and can not be used directly, but needs to go through further degluing and impurity removal. The processed Hang Leno has comfortable handling. After dyeing, rinsing and ironing in line with different needs, the fine Hang Leno can be produced. At present, most of the Hang Leno are plain type. Therefore, decorative processes such as printing and dyeing can be carried out on the finished products to improve the appearance.

五、现代丝绸罗的加工织造/Weaving Process of Modern Silk Leno

丝绸罗织物的传统加工织造主要采用手工木织机，例如周家明自制了四经绞罗的手工木织机，再现了四经绞罗织物的织造，如图2-12所示。该木织机主要由开口（绞综和素综）、投梭、打纬（木筘）、送经（经轴）、卷取（织轴）等五大主要机构组成。该木织机没有花楼机构，只能生产素罗织物。

The traditional weaving process of silk leno mainly adopts the hand knitting machine. For example, Zhou Jiaming invented his own loom for making four-warp twist ribs, as shown in Figure 2-12. The wood loom mainly contains five functions: opening（twisted heald and plain heald）, picking, weft beating（wood cop）, lifting（warp beam）and coiling（weaving shaft）. Since the

图2-12　周家明和四经绞罗手工木织机
Zhou Jiaming & His Wooden Loom for Making Four-warp Twist Ribs

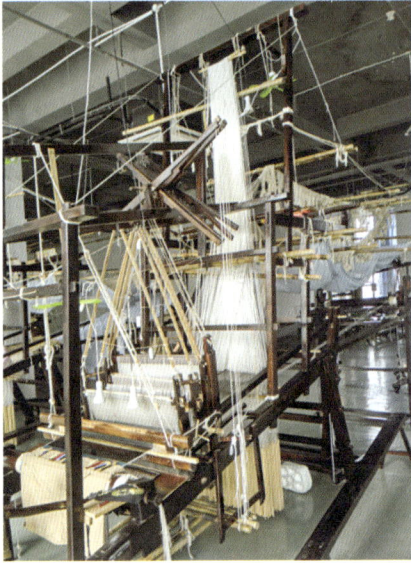

图2-13　含有花楼机构的提花罗木织机
Leno Loom with Jacquard Device

wood loom doesn't have jacquard mechanism, it can only produce plain leno.

胡德银等曾在苏州汪永亨丝绸科技文化有限公司工作，为了开发新的丝绸罗织物产品，他们在传统的素罗织机上增加了花楼机构，设计与制作了提花罗织机，如图2-13所示。通过改变织物的花本，可以生产具有一定图案结构的提花罗织物，便于开发丝绸新产品。

Hu Deyin, as well as other workers, had worked in Wang Yongheng Silk Technology and Culture Company. To develop a new type of silk leno, they installed a jacquard device into a traditional loom designed for plain leno. Thus, a new jacquard loom had been designed, as shown in Figure 2-13. By changing the template, the jacquard loom can produce different patterned jacquard leno, which is conducive for developing new silk products.

传统的木织机结构复杂，生产效率低，例如含有花楼机构的提花罗木织机需要两个人才能完成织造，制约了丝绸罗织物的快速生产。后来通过对有梭织机进行相应的改造，即增加提花开口机构和绞综，实现了丝绸罗织物的铁制织机织造，提高了织造效率。例如，苏州市锦达丝绸有限公司织造提花罗织物采用的就是有梭织机，如图2-14所示。

The structure of a traditional loom is complicated with low efficiency. For example, the leno loom with jacquard device needs two people to complete weaving, which restricts the rapid production of silk leno. Afterwards, technicians upgraded the shuttle loom, improving the mechanism and twisted heald. It enabled the iron loom to weave silk leno and improved production efficiency. For example, Suzhou Jinda Silk Company adopts a shuttle loom to weave jacquard leno, as shown in Figure 2-14.

图2-14　织造提花罗用的有梭织机
Shuttle Loom for Jacquard Leno

○ 第三章

罗组织的结构
The Structure of Leno

第一节	罗组织的概念和形成过程/The Concept and Formation Process of Leno Weaves

一、罗组织的基本概念/The Basic Concepts of Leno Weaves

　　罗组织是纱罗组织的一种，一般的纱罗组织由罗组织和纱组织共同构成，而纱组织和罗组织统称为纱罗组织。纱罗组织由绞经与地经组成，绞经与地经形成一个绞组，在与纬纱交织时，地经位置不动，同一绞组的绞经有时在其左侧，有时在其右侧，当绞经从地经的一侧转到另一侧时，形成一次扭绞。同一绞组的绞经每改变一次左右位置，形成一次扭绞时，织入1根纬纱或共口的数根纬纱，形成的组织称为纱组织，如图3-1（a）和图3-1（b）所示。同一绞组的绞经每改变一次左右位置，形成一次扭绞时，织入3根或3根以上奇数纬纱，形成的组织称为罗组织，如图3-1（c）和图3-1（d）所示，其中图3-1（c）为三梭罗，图3-1（d）为五梭罗。

　　Leno weave is a kind of gauze. Ordinary leno weaves are composed of leno weaves and gauze weaves. The gauze and leno weaves consist of strand warp and ground warp, and those two warps form a leno heald group. When the strand warp interwines with the ground warp, the latter does not change position. The strand warp of the same group is sometimes on the left side and sometimes on the right side. When the ground warp passes from one side of the ground warp to the other, a twist is formed. As twists forming, strand warp with one or more than one ground warps at one position is called gauze and leno weaves, as shown in Figures 3-1 (a) and (b) . A twist with three or more odd numbers of ground warps is called leno weaves, as shown in Figures 3-1 (c) and (d) , among which the Figure 3-1 (c) is three twist leno, and Figure 3-1 (d) is five twist leno.

　　形成一个纱孔所需的绞经与地经称为一个绞组，下面以三梭罗为例来说明罗组织的几种绞组。当绞经∶地经＝1∶1，即一个绞组由一根绞经和一根地经组成，称为一绞一，如图3-2（a）所示；当绞经∶地经＝1∶2，即一个绞组由一根绞经和两根地经组成，称为一绞

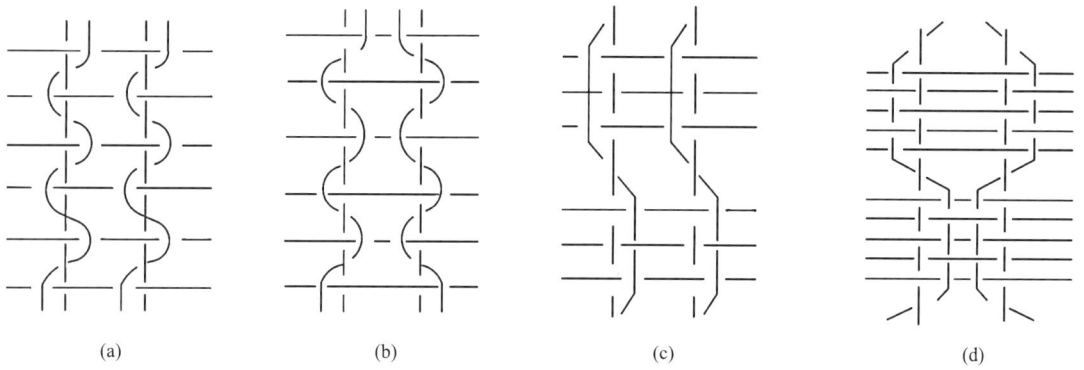

(a) (b) (c) (d)

图3-1 纱罗组织结构图
Structure of Gauze and Leno Weaves

二，如图3-2（b）所示；当绞经：地经＝2：2，即一个绞组由两根绞经和两根地经组成，称为二绞二，如图3-2（c）所示。

The strand warp and ground warp required to form a yarn hole is called a twisted group. Taking three twist leno as an example, there are several types of twisted groups. When strand warp：ground warp = 1：1, that is, a twisted group consists of one strand warp and one ground warp, called one twist one, as shown in Figure 3-2（a）; when the strand warp：ground warp = 1：2, that is, a group of twisted composed of one strand warp and two ground warps, called one twist two, as shown in Figure 3-2（b）; when strand warp：ground warp = 2：2, that is, a twisted group consists of two strand warps and two ground warps, called two twist two, as shown in Figure 3-2（c）.

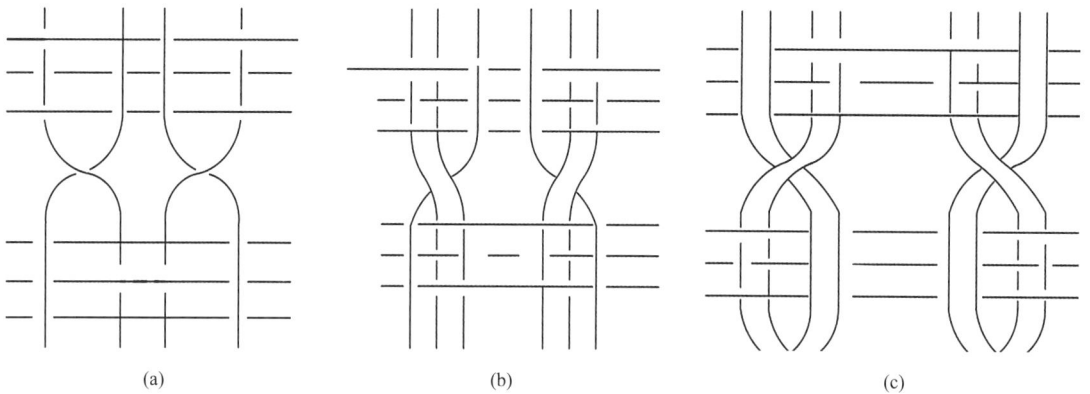

(a) (b) (c)

图3-2 罗组织常用的三种绞组
Three Types of Twisted Groups

在各绞组中，绞经与地经绞转方向均一致的罗组织称为一顺绞。相邻两个绞组内，绞经与地经绞转方向相对称的罗组织称为对称绞。在其他条件相同的情况下，对称绞所形成的纱孔比一顺绞所形成的纱孔清晰。

In each group, the leno weaves with the same direction of strand warp and ground warp is called a longitudinal twist. In two adjacent groups, the leno weaves with symmetrical strand warp and ground warp is called a symmetrical twist. All other things being equal, the yarn hole formed by a symmetrical twist is clearer than that of a longitudinal twist.

二、罗组织的形成 /The Formation of Leno Weaves

（一）绞综结构 /Structure of the Twisted Heald

罗组织依靠一种特殊的绞综来实现绞经与地经的相互扭绞。绞综有线制绞综和金属绞综。线制绞综结构简单，容易制作；金属绞综结构复杂，制作成本较高，但应用方便。目前制作结构简单、密度较低的罗织物时主要使用金属绞综，但在制作花型复杂或密度较大的罗织物时，仍需采用线制绞综。

The intertwined strand warp and ground warp of leno weaves is accomplished by a special twisted heald. There are two types of twisted healds, one is made of wire, and the other is made of metal. Compared with the simple structure and production process of wire twisted heald, the metallic one has a complex structure and costs a lot during production, but it is easy to use. At present, metallic twisted healds are mainly used for the production of simple, low–density leno fabrics, while wire twisted healds are still required for the production of complex patterns and denser leno fabrics.

1. 线制绞综 /Wire Twisted Heald

线制绞综由基综和半综联合而成。目前仍在使用的基综有两种，一种是普通金属综丝，其使用寿命较长；另一种是线制综丝，用比较细的尼龙线穿过玻璃或铜的目销子的上下孔眼，目销子中间孔眼穿有半综。线制基综的使用寿命较短。

Wire twisted healds consist of a base heald and a semi–heald. At present, there are two kinds of base healds in production, one is common metallic wire, which has a long service life, and the other is a kind of wire base with thin nylon thread passing through the upper and lower holes of the glass or copper pin, which is equipped with harness wire. The service life of the wire base composite is shorter.

半综为尼龙线制成的环圈，有上半综和下半综之分。半综上端穿过基综综眼，下端固定在一根棒上，由弹簧控制，称为下半综，如图3-3（a）和图3-3（b）所示。若将半综的上端固定，下端穿过基综综眼，则称为上半综，如图3-3（c）所示。下半综用于上开梭口和中央闭合梭口的织机，使用较多。上半综用于下开梭口的织机，使用较少。半综按环圈头的伸向不同，又有左半综和右半综之分。凡半综环圈头伸向基综左侧，即绞经线应位于基综的左侧称为左半综，如图3-3（b）所示；半综环圈头伸向基综右侧，即绞经线应位于基综的右侧称为右半综，如图3-3（a）和图3-3（c）所示。上机前，半综杆位于基综的前方。

The semi–heals are rings made of nylon wire, which can be divided into the upper half heald and the lower half heald. The upper end of the semi–heald passes through the eye of the base heald, and the lower end is fixed on a rod controlled by a spring, which is called the lower half heald, as shown in Figures 3–3（a）and（b）. Inversely, the upper half heald refers to the heald with the upper end of the semi–heald fixed on a rod, as shown in Figure 3–3（c）. The lower half heald is used for the loom with upper opening shed and central closed shed, and is frequently used. The upper half is used in the loom with a lower opening shed, and it is less used. The semi-heald can be classified into the left half heald and the right half heald according to the extension direction of the ring head. Where the ring head extends to the left of the base heald, i.e., the stand warp should be located to the left of base heald is called the left half heald, as shown in Figure 3-3 (b) ; Where the ring head extends to the right of the base heald, i.e., the stand warp should be located to the right of the base heald is called the right half heald, as shown in Figures 3-3 (a) and (c) . Before starting the loom, the semi-heald rod is located in front of the base coal.

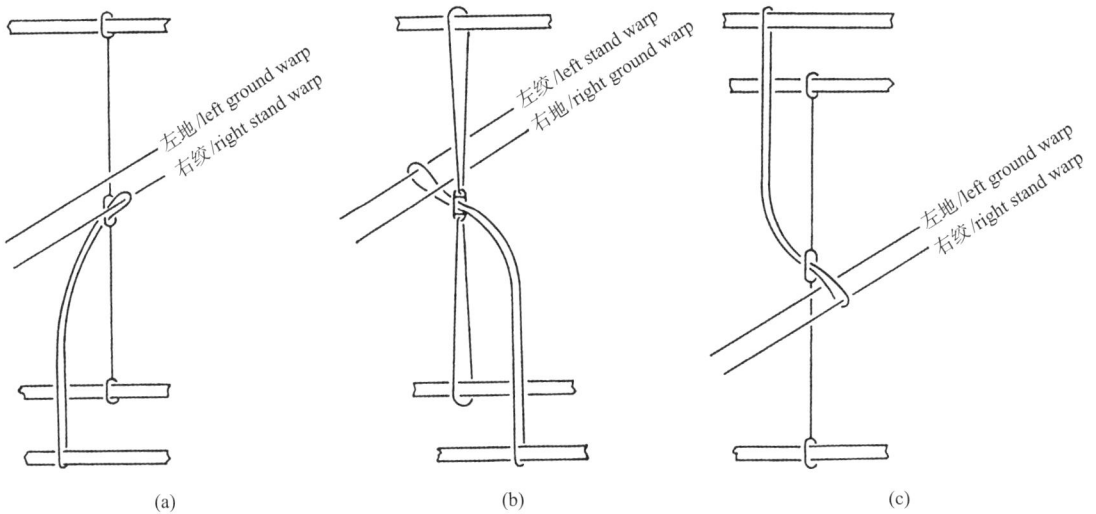

(a)　　　　　　　　　(b)　　　　　　　　　(c)

图3-3　线制绞综
Wire Twisted Heald

2. 金属绞综/Metallic Twisted Heald

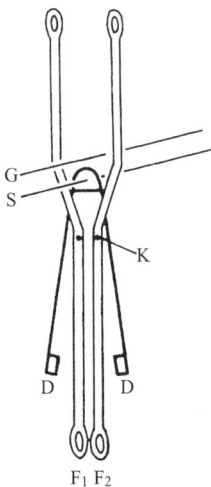

图3-4　金属绞综
Metallic Twisted Heald

金属绞综由左、右两根基综F$_1$、F$_2$和一根半综D组成，如图3-4所示。每根基综由两片扁平的钢质薄片组成，并由中部的焊接点K将两薄钢片连成一体。半综D骑跨在两基综之间，半综的每一支脚伸在基综上部的两薄片之间，并由基综的焊点K拖持。这样，无论哪一片基综上升，半综都能随之上升，以改变绞经S与地经G的相对位置。图3-4中半综的两只脚朝下，称为下半综。若将半综的两只脚朝上，则称为上半综。织造时通常采用下半综，以形成上开梭口。

The metallic twisted heald is made of two basic heddles F$_1$, F$_2$ and one semi-heald D, as shown in Figure 3-4. Each base heald is composed of two flat steel sheets connected by the welding point K in the middle. The semi-heald D is spanning two base healds, with each end extending to the middle of two healds and holding by the welding point K. In this way, no matter which base heald rises, the semi-heald can rise accordingly to change the relative position of the stand warp S and the ground warp G. Lower half heald with two downward legs is shown in Figure 3-4. Semi-heald with upward legs is called upper half heald. The lower half heald is usually adopted to form the upper shed.

（二）穿综方法/Methods of Healding

罗组织的穿综方法与一般织物不同，需要经过两步。第一步将绞经与地经分别穿入位于机后的普通综丝，其中穿绞经的综称为后综，穿地经的综称为地综；第二步将每一绞组内的绞经穿入半综综眼，将地经从同组绞经穿入的半综的两基综丝间穿过。

Different from the general fabric, the healding method of leno consists of two processes. The first step is to thread the strand warp and ground warp separately through the heddle at the back of the loom. Those heddles threaded by the stand warp are called the latter healds, and those with ground warp are called ground healds. The second step is to thread the strand warp of each twisted group through the eye of the semi-heald and at the same time, thread the ground warp through the middle of the threaded semi-heald.

在穿综时确定同一绞组的绞经和地经的相互位置。由于多臂机与提花机为上开梭口或中央闭合梭口，需采用下半综制织。下面以下半综一绞一为例说明其具体穿法。

The position of intertwined warps is determined by the time of the healding. Because the upper opening shed or central closed shed of the dobbies and jacquard machine, the lower half heald is needed. Take one twist one of lower half heald as an example to illustrate the specific method of healding.

1. 右穿法/Right Crossing Method

从机前看，绞经从地经的右侧穿入半综综眼。从左至右，机上经纱第1根为地经，第2根为绞经。图3-5（a）为线制绞综的右穿法示意图。采用右半综制织，绞经穿入后综时，位于地经右侧，绞综提升使绞经绕过地经的下方，从地经的右侧绞转到左侧。图3-5（b）为金属绞综的右穿法示意图。绞经S穿入后综B时位于地经右侧，然后自基综F_2的左侧和基综F_1的右侧之间穿入半综D的孔眼，地经G穿入地综A后，再与绞经同样的方位穿过两基综之间。结果是前基综F_1提升，使绞经S绕过地经下方，从地经的右侧绞转到左侧。

From the front of the loom, the strand warp passes through the eye of the semi-heald from the right side of the ground warp. From left to right, the gauze on the loom is ground warp firstly and strand warp secondly. Figure 3-5 (a) is a schematic diagram of right crossing method of wire twisted heald. By adopting the right half heald, the strand warp is on the right side of the ground warp after crossing the latter heald. The lifted twisted heald makes the strand warp pass around the bottom of the ground warp, and then turn to its left side from its right side. Figure 3-5 (b) is a schematic diagram of right crossing method of metallic twisted heald. The strand warp S is located on the right side of the ground warp when it passes through the latter heald B, and then threads the eye of semi-heald D from the place between the left side of base-heald F_2 and the right side of the base-heald F_1. After passing the ground heald, the ground warp G crosses the middle place of two base-healds in the same direction as the strand warp. The result is that the lifting of F_1 of front base-heald makes the strand warp S pass around the bottom of ground warp, and then twist from the right side of the ground warp to its left side.

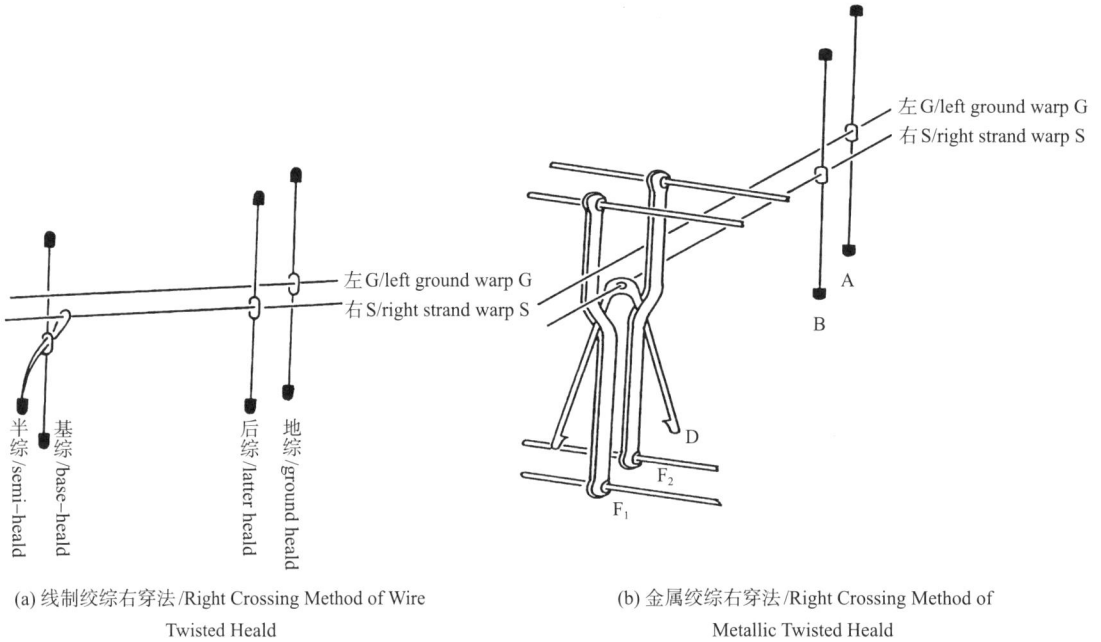

左 G/left ground warp G
右 S/right strand warp S

左 G/left ground warp G
右 S/right strand warp S

半综/semi-heald
基综/base-heald
后综/latter heald
地综/ground heald

A
B
D
F₂
F₁

(a) 线制绞综右穿法/Right Crossing Method of Wire Twisted Heald

(b) 金属绞综右穿法/Right Crossing Method of Metallic Twisted Heald

图 3-5 右穿法
Right Crossing Method

2. 左穿法/Left Crossing Method

从机前看，绞经从地经的左侧穿入半综综眼。从左至右，机上经纱第 1 根为绞经，第 2 根为地经。图 3-6（a）为线制绞综的左穿法示意图。采用左半综制织，绞经穿入后综时，位于地经左侧，绞综提升使绞经绕过地经的下方，从地经的左侧绞转到右侧。图 3-6（b）为金属绞综的左穿法示意图。绞经 S 穿入后综 B 时位于地经左侧，然后自基综 F₂ 的右侧和基综 F₁ 的左侧之间穿入半综 D 的孔眼，地经 G 穿入地综 A 后，再与绞经同样的方位穿过两基综之间。结果是前基综 F₁ 提升，使绞经 S 绕过地经下方，从地经的左侧绞转到右侧。

From the front of the loom, the strand warp passes through the eye of the semi-heald from the left side of the ground warp. From left to right, the gauze on the loom is strand warp firstly and ground warp secondly. Figure 3-6 (a) is a schematic diagram of left crossing method of wire twisted heald. By adopting the left half heald, the strand warp is on the left side of the ground warp after crossing the latter heald. The lifted twisted heald makes the strand warp pass around the bottom of the ground warp, and then turn to its right side from its left side. Figure 3-6 (b) is a schematic diagram of left crossing method of metallic twisted heald. The strand warp S is located on the left side of the ground warp when it passes through the latter heald B, and then threads the eye of semi-heald D from the place between the right side of base-heald F₂ and the left side of the base-heald F₁. After passing the ground heald A, the ground warp G crosses the middle place of two base-healds in the same direction as the strand warp. The result is that the lifting of F₁ of front base-heald makes the strand warp S pass around the bottom of the ground warp, and then twist from the left side of the ground warp to its right side.

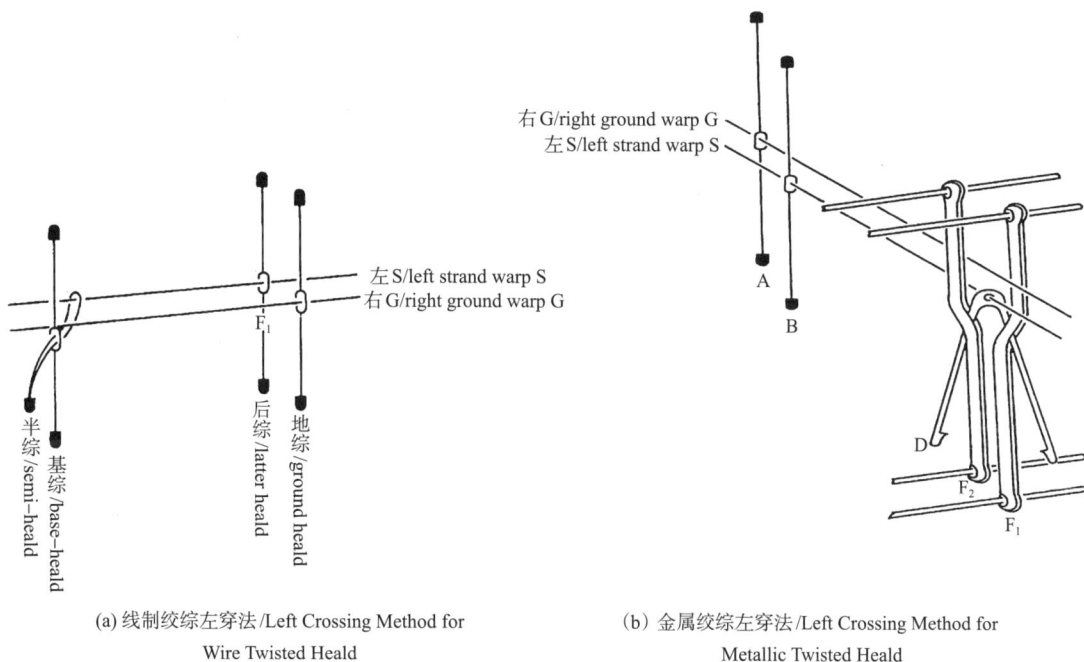

（a）线制绞综左穿法/Left Crossing Method for
Wire Twisted Heald
　　　　（b）金属绞综左穿法/Left Crossing Method for
Metallic Twisted Heald

图3-6　左穿法
Left Crossing Method

上机时，若金属绞综采用单一的左穿法或右穿法，只能获得一顺绞。若要获得对称绞，应左穿法与右穿法混合使用；线制绞综同理。

The single stands in sequence can be obtained if metallic twisted heald only adopted left crossing method. To obtain symmetrical strands, both the left and right crossing methods should be mixed, so does the wire twisted heald.

（三）梭口/Shed

织制罗织物时，根据开口时绞经与地经的相对位置不同，梭口可分为绞转梭口、开放梭口和普通梭口三种。由于线制绞综与金属绞综的结构不一样，形成上述三种梭口的提综情况也有差别，下面分别阐述两种绞综的三种梭口。

According to the relative position of strand warp and ground warp when shedding, the shed of leno fabric can be divided into three types: twisted shed, open shed and general shed. Since the structure of wire twisted heald is different from that of metallic twisted heald, the shedding condition of the above three kinds is also different. Three sheds of two types of twisted heald are described as follows.

1. 线制绞综的三种梭口/Three Types of Sheds for Wire Twisted Heald

（1）绞转梭口。后综与地综静止不动，由基综和半综（统称为绞综）提升所形成的梭口称绞转梭口，如图3-7（a）所示。图中采用右半综右穿法，即原上机位置为绞经在地经右侧。当织入第4纬时，绞综提升使绞经S从地经G下面转绕到地经右侧升起，形成梭口的上层；地经G不动，形成梭口的下层。

Twisted Shed. Putting the latter heald and the ground warp at standstill, the twisted shed is formed by lifting base–heald and semi–heald (collectively refer to twisted heald), as shown in Figure 3–7 (a). In the figure, the right crossing method is adopted, that is, the original upper machine position is the strand warp on the right side of the ground warp. When the 4th weft is woven in, the base–heald and semi–heald are lifted so that the strand warp S turns from the ground warp G to the right side of the ground warp and rises to form the upper layer of the shed, and the ground warp G does not move to form the lower layer of the shed.

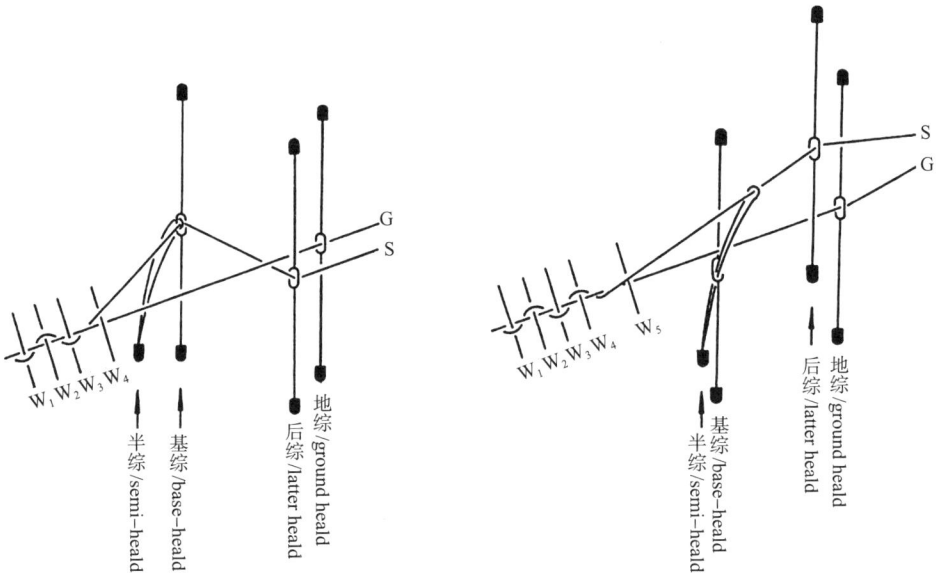

(a) 绞转梭口/Twisted Shed

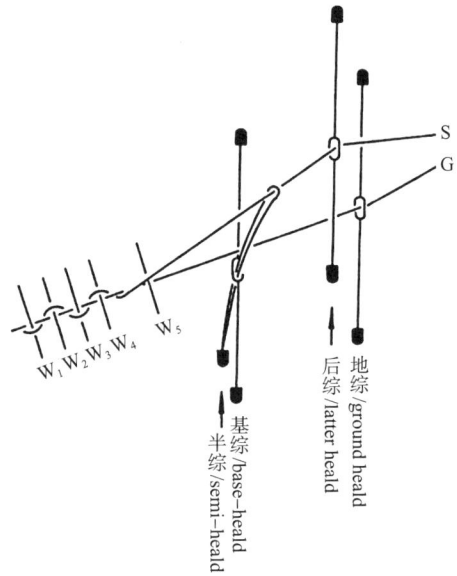

(b) 开放梭口/Open Shed

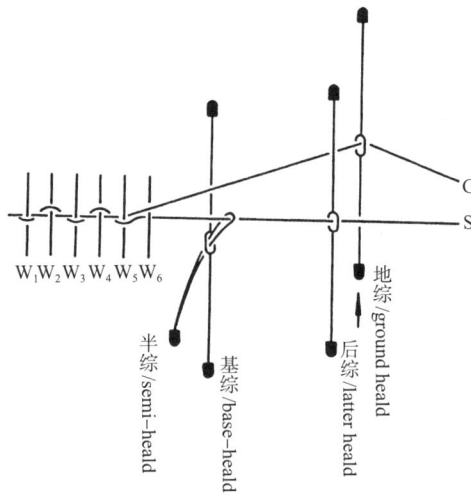

(c) 普通梭口/General Shed

图3-7　线制绞综的三种梭口
Three Types of Sheds for Wire Twisted Heald

（2）开放梭口。地综与基综静止不动，由后综和半综提升所形成的梭口称开放梭口，如图3-7（b）所示。当织入第5纬时，后综、半综提升使绞转在地经G左侧的绞经S仍回到地经的右侧（原来上机位置）上升，形成梭口的上层；地经G不动，形成梭口的下层。

Open shed.Putting the ground heald and the base heald at standstill, the open shed is formed by lifting the latter heald and semi-heald, as shown in Figure 3-7（b）. When the 5th weft is woven in, the latter heald and semi-heald are lifted so that the strand warp S on the left side of the ground warp still returns to the right side of the ground warp（the original position on the machine）and rises, forming the upper layer of the shed and the ground warp G does not move, forming the lower layer of the shed.

（3）普通梭口。后综与绞综静止不动，由地综提升所形成的梭口称普通梭口，如图3-7（c）所示。当织入第6纬时，地经G由地综带动上升，形成梭口上层，绞经S形成梭口下层。绞经与地经的相对位置与前一纬相同，绞经仍在地经的右侧，相互没有扭绞。

General shed. Keeping the latter heald and the twisted heald at standstill, the general shed is formed by lifting the ground heald, as shown in Figure 3-7（c）. When the 6th weft is woven in, the ground warp G is driven up by the ground heald to form the upper layer of the shed, and the warp S is twisted to form the lower layer of the shed. The relative position of the strand warp and the ground warp is the same as the previous weft, with the strand warp on the right side of the ground warp without twisting with each other.

2. 金属绞综的三种梭口/Three Types of Sheds for Metallic Twisted Heald

以常用的右穿法（左绞穿法）为例，说明三种梭口的形成。综平时绞经S位于地经G的右侧。

The commonly used right crossing method（left twisted threading method）is taken as an example to illustrate the forming of three kinds of shed. The strand warp is placed on the right side of the ground warp G during the healding.

（1）绞转梭口。如图3-8（a）所示，由基综F_1和半综D上升，使绞经S从地经G下面扭转到地经左侧升起形成的梭口。

Twisted shed. As shown in Figure 3-8（a）, the twisted shed is formed by lifting base heald F_1 and semi-heald D to let strand warp S twist from the bottom of ground warp G to its left side.

（2）开放梭口。如图3-8（b）所示，由基综F_2和半综D上升，同时后综也提升，使绞经S仍回到地经G的右侧（原来上机位置）升起形成的梭口。

Open shed. As shown in Figure 3-8（b）, the open shed is formed by lifting the base-heald F_2, semi-heald D, as well as latter heald in order to make strand warp return to the right side of the ground warp D（the original position）.

（3）普通梭口。如图3-8（c）所示，由地综提升，使地经G升起形成梭口，绞经、地经相对位置同前一纬梭口。织制绞纱组织时，只要交替地使用绞转梭口与开放梭口，使绞经时而在地经的左侧，时而在地经右侧，相互扭绞而形成纱孔。地综不运动，地经始终位于梭口下层，而半综每一梭都要上升，或随着基综上升，或随着后综上升，而不可能单独提升形成梭口。

General shed. As shown in Figure 3-8 (c) , lifting the ground warp caused by lifted ground heald forms the general shed, with strand and the ground warp are the same as that of the previous weft shed. When weaving twisted yarn, as long as the twisted shed and open shed are used alternately to make the strand warp sometimes on the left side of the ground warp and sometimes on its right side, and to twist each other to form yarn holes. With the ground warp being stationary, the ground warp is always located at the lower layer of the shed, while each shuttle of the semi-heald must rise, either with the rise of the base-heald or with the rise of the latter heald. It can not be lifted alone to form a shed.

织制罗组织时，地经也要提升，因此应同时采用三种梭口。例如，织三梭罗，梭口顺序为：开放梭口—普通梭口—开放梭口，绞转梭口—普通梭口—绞转梭口；织五梭罗，梭口顺序为：开放梭口—普通梭口—开放梭口—普通梭口—开放梭口，绞转梭口—普通梭口—绞转梭口—普通梭口—绞转梭口。

When weaving leno, the ground warps should also be raised, so three kinds of sheds should be used. If weaving three shuttle gauze, the order of sheds is: open—general—open, twisted—general—twisted, and the order of the sheds when weaving five shuttle gauze is: open shed—general shed—open shed—general shed—open shed, twisted shed—general shed—twisted shed—general shed—twisted shed.

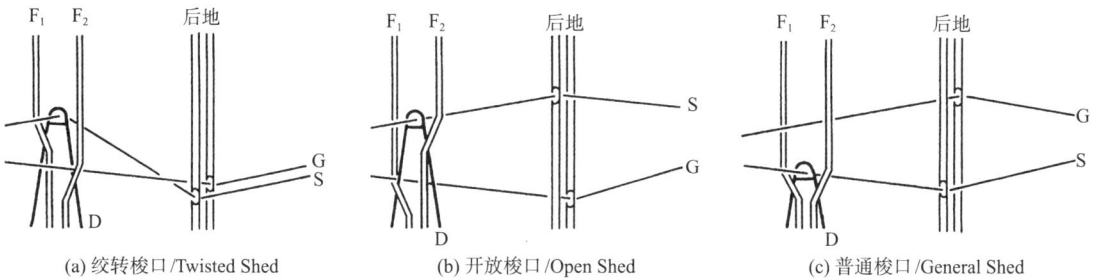

(a) 绞转梭口 /Twisted Shed (b) 开放梭口 /Open Shed (c) 普通梭口 /General Shed

图 3-8　金属绞综的三种梭口
Three Types of Sheds for Metallic Twisted Heald

第二节　罗组织的上机 /The Looming Plans of Leno Weaves

一、上机图的绘制 /Graphic Design for the Looming Plans of Leno Weaves

罗组织由绞经和地经扭绞形成，又受到线制绞综或金属绞综等特殊条件的影响，因此罗组织的上机图与一般组织有所区别。图 3-9（a）为线制绞综左穿法上机图；图 3-9（b）为金属绞综左穿法上机图；图 3-9（c）为线制绞综右穿法上机图；图 3-9（d）为金属绞综右穿法上机图。

The leno fabric is formed by twisted stand warps and ground warps, and influenced by special conditions such as wire twisted heald or metallic twisted heald. Therefore, the graphic design of leno weaves is different from that of general structure. Figure 3-9 (a) is the graphic design of left crossing method for wire twisted heald; Figure 3-9 (b) is the graphic design of left crossing method for metallic twisted heald; Figure 3-9 (c) is the graphic design of right crossing method for wire twisted heald; Figure 3-9 (d) is the graphic design of right crossing method for metallic twisted heald.

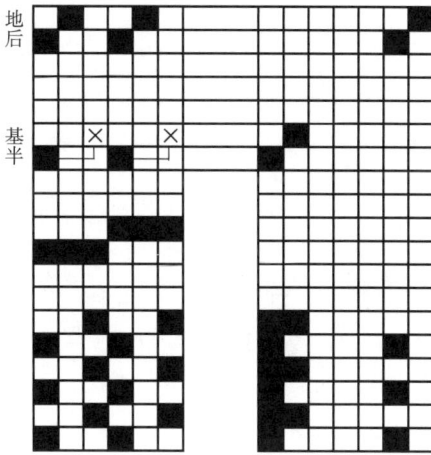

(a) 线制绞综左穿法 /Left Crossing Method for Wire Twisted Heald

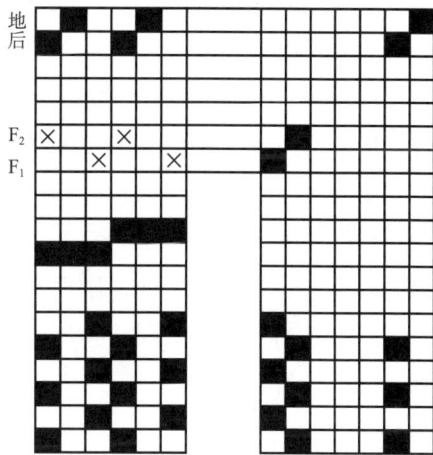

(b) 金属绞综左穿法 /Left Crossing Method for Metallic Twisted Heald

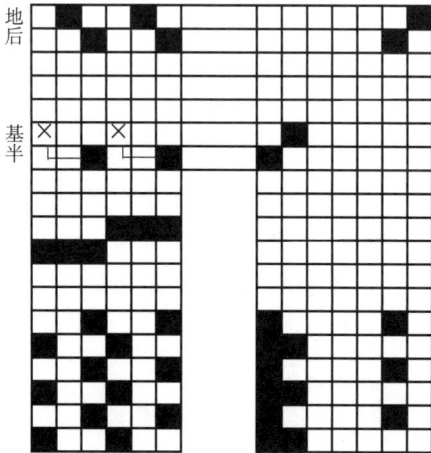

(c) 线制绞综右穿法 /Right Crossing Method for Wire Twisted Heald

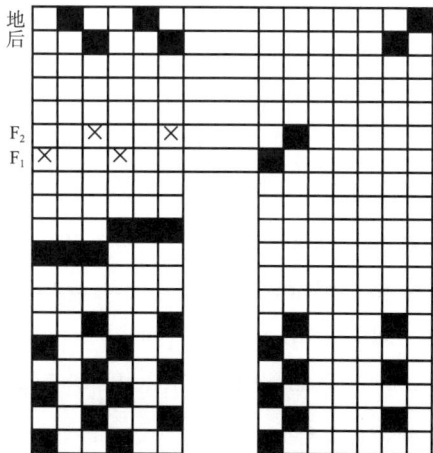

(d) 金属绞综右穿法 /Right Crossing Method for Metallic Twisted Heald

图3-9　罗组织上机图
Graphic Design for Leno Weave

1. 组织图/Graphics of Leno Weave

由于罗组织的绞经时而在地经的右侧，时而在地经的左侧，所以绘制组织图时，绞经在地经的左右两侧各占一个综列，并标以同样的序号。每个绞组的经纱占几个综列，需根据绞组的结构而定。如一绞二，表示一个绞组有1根绞经和2根地经，则一个绞组需占用四纵行；二绞二则表示一个绞组有2根绞经和2根地经，则一个绞组需占用六纵行。在组织图（图3-9）中，符号"■"表示绞经的浮点，符号"⊠"表示地经的浮点。

Since the strand warp of leno weaves is sometimes on the right side of the ground warp and sometimes on the left side of the ground warp. When drawing the leno weave, the strand warps separately occupy a row on both the left and right sides of the ground warps, and are marked with the same serial number. The numbers of rows occupied by warp of each twisted group depend on its structure. For example, one twist two, which means that a twisted group has one strand warp and two ground warps, occupies four columns, while two twist two, which means that a twisted group has two strand warps and two ground warps, occupies six columns. In the weaving chart (Figure 3-9), the symbol "■" represents the floating point of the strand warp and the symbol "⊠" represents the floating point of the ground warp.

2. 穿筘图/Graphics of Denting

用两横行表示，连续涂绘的方格代表该绞组内的经纱与地经穿入同一筘齿，并不代表经纱根数。如图3-9所示，上机时，横向连续涂绘的三格仅表示1根绞经和1根地经穿入同一个筘齿中。

The continuous painted square, expressed by two horizontal lines, represents that the strand warp and the ground warp within a twisted group thread one denting, rather than representing the number of strand warp. As shown in Figure 3-9, the three spaces painted horizontally and continuously indicate only one strand warp and one ground warp passing through the same denting when weaving.

3. 穿综图/Graphics of Heald

每一横行代表一片综框，符号"■"表示绞经或地经穿入该片综，"⊠"表示基综的位置。右穿法为左侧基综在前，右侧基综在后；左穿法为右侧基综在前，左侧基综在后。

Each horizontal line represents a heald frame. The symbol "■" indicates the strand warp or ground warp crossing into the heald, while "⊠" refers to the position of the base heald. Right crossing method: the left base heald is in front of the right base heald; left crossing method: the right base heald is in front of the left base heald.

4. 纹板图/Lifting Plan

金属绞综起绞转梭口时，仅前基综F_1提升，后基综F_2不提升，而在起开放梭口时，后综与后基综F_2同时提升；线制绞综的基综与半综提升形成起绞转梭口，后综与半综提升形成开放梭口，为此，半综在起绞转梭口和开放梭口时均需提升。

As the metallic twisted heald forms twisted shed, only the front base heald F_1 is rising while the latter one does not change. As it forms open shed, both the latter heald and the latter base heald F_2 are lifting simultaneously. The base heald of wire twisted heald and semi-heald are lifted to form a twisted shed, and the latter heald and half semi-heald are lifted to form an open shed. Therefore, the half heald needs to be lifted when both the twisting shed and the open shed are lifted.

二、上机要点/The Key Point of Weaving

（1）罗织物由于绞经和地经的运动规律不同，两者的收缩率不同，有时差异很大，必要时采用两个织轴织造，尽可能使用一个织轴织造。

Because of the different movement rules of strand warps and ground warps, the shrinkage of the two is different and sometimes the difference is huge. Use two live axles if necessary, and use one live axle as much as possible.

（2）每一绞组必须穿入同一筘齿，否则打纬时将切断经纱，进而无法织造。为了加大纱孔，突出扭绞风格，采用空筘法或花式筘穿法。

Each twisted group must be pierced with the same denting, otherwise, the strand warp will be cut off during beating, and it can not be woven. In order to enlarge the yarn hole and highlight the twist style, empty denting or fancy denting is adopted.

（3）为了保证开口清晰度，减少断经，则绞综在前面，其他组织在中间，后综与地综在最后。绞综与地综的间隔以3～5片综框为宜。对采用绞经、地经合轴织造的品种，这个距离尤其重要。

In order to ensure the clarity of opening and reduce the broken warps, the twisted heald is in the front, other organizations are in the middle, and the latter heald and ground heald are at the ends. The interval between the twisted heald and the ground heald should be 3−5 heald frames. This distance is particularly important for the varieties of weaving with the twisted heald and the ground heald.

（4）采用金属绞综织制纱罗织物，综平时应使地经稍高于半综的顶部，以便绞经在地经之下顺利绞转。采用线制绞综织制纱罗织物，综平时应使绞综综眼低于地综综眼，半综环圈头伸出基综综眼2～3mm，以便绞经在地经之下顺利地左右绞转，形成清晰梭口。

As for the leno weave made of metallic twisted healds, the ground warp shall be slightly higher than the top of the semi-heald when weaving, so that the strand warp can be twisted smoothly under the ground warp. The leno weave is made of wire twisted heddles, and the twisted heddle eyelet should be lower than the ground heddle eyelet when weaving, and the semi-heald loop head should extend 2−3mm from the base heddle eyelet so that the strand warp can be smoothly twisted under the ground warp to form a clear shed.

| 第三节 | 罗组织的应用实例/The Application Examples of Leno Weaves |

罗组织结构与其他织物组织结构相比，典型特征是存在扭绞，通过绞经和地经的扭转形成一个个罗孔，在织物表面呈现出罗纹效果。

Compared with other fabrics, the typical feature of leno weave is that there are twists, which form holes by twisting the strand warps and the ground warps, to present the rib on the surface of the fabric.

根据织物表面罗纹排列的形状，罗纹呈横条的称为横罗，如图3-10（a）所示；罗纹呈直条的称为直罗，如图3-10（b）所示。

According to the shape of the rib arrangement on the surface of the fabric, the rib with horizontal strip is called horizontal leno, as shown in Figure 3-10（a）; the rib with vertical strip is called vertical leno, as shown in Figure 3-10（b）.

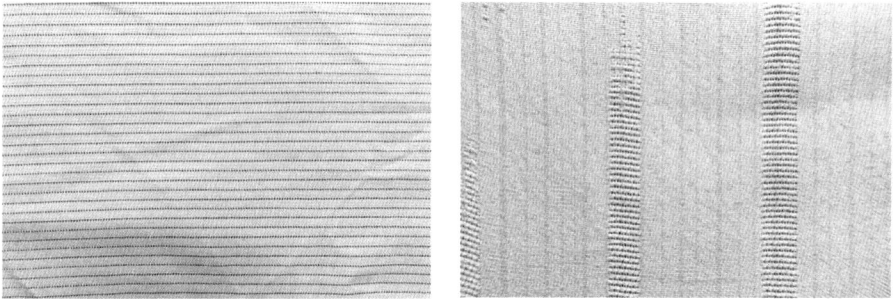

(a) 横罗 /Horizontal Leno   (b) 直罗 /Vertical Leno

图3-10　横罗和直罗织物
Horizontal and Vertical Leno

根据是否采用提花工艺可分为素罗和花罗。图3-10（a）面料中没有提花，称为素罗；图3-11面料中纬纱为十七的横罗上加入提花组织，称为花罗。

According to whether jacquard technology is adopted, it can be divided into plain leno and patterned leno. In Figure 3-10（a）, fabric with no jacquard can be called plain leno, while Figure 3-11 shows that jacquard tissue is added to the horizontal roller with weft 17, which is called patterned leno.

根据织物组织结构可分为二经绞罗、三经绞罗、四经绞罗、四经互绞几何纹花罗、十经互绞花罗和多经（十二经以上）互绞花罗等。市场上常见的品种有杭罗、纹罗、帘锦罗、四经绞罗等，还有一些新品种，如变形罗、方格罗、曲经绞罗［图3-12（a）］、双面异色花罗［图3-12（b）］等。

In terms of the fabric weave, there are twisted leno with two warps, twisted leno with three warps, twisted leno with four warps, twisted leno with four warps and geometric patterns, patterned twisted leno with ten warps, and patterned twisted leno with more than twelve warps. There are some common varieties in the market, such as Hang Leno, Patterned Leno, Lianjin Leno, twisted leno with four warps, etc., and some new

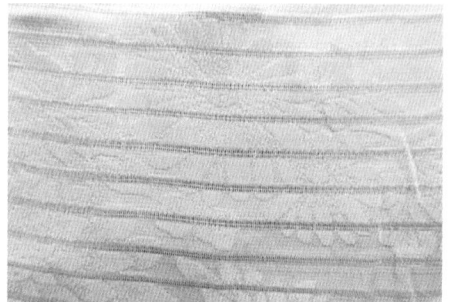

图3-11　花罗
Patterned Leno

varieties such as deformed leno, square leno, curving warp twisted leno［Figure 3-12（a）］, and leno with double-sided heterochromatic flower［Figure 3-12（b）］.

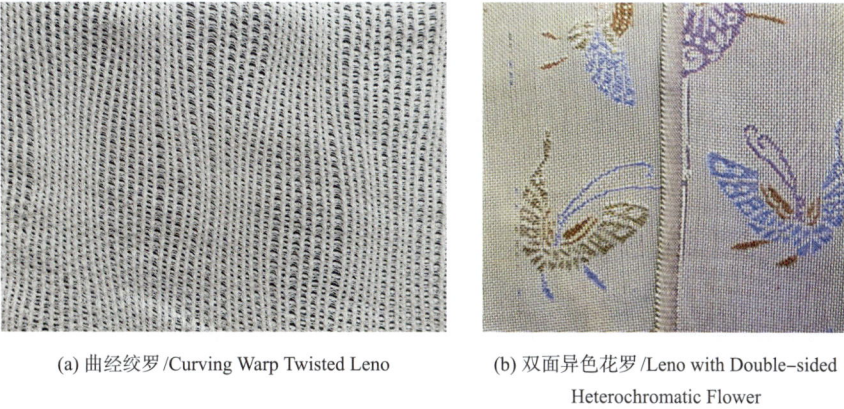

(a) 曲经绞罗 /Curving Warp Twisted Leno

(b) 双面异色花罗 /Leno with Double-sided Heterochromatic Flower

图3-12 罗织物新品种
New Varieties of Leno Weaves

这些变化较多的罗织物在织造时基本都采用线制绞综来完成。罗织物面料紧密结实，身骨平挺爽滑，透气性好，表面布满各种不同形态比纱孔还要稀疏的孔眼。也有高密度孔眼的罗织物，但有着明显的凹凸感和立体感，适合制作窗纱、屏风、服装和内饰等。

These leno fabrics with more changes are basically finished by wire twisted healds. Leno fabrics, with good air permeability, are compact, smooth flat, and covered with various kinds of holes which are more sparse than the yarn holes on the surface. They also have special high-density roller holes with obvious bump and three-dimensional shape, which are suitable for making window yarn, screen, clothing, interior decoration, etc.

第四章

罗的传承与保护
The Inheritance and Protection of Leno

第一节　罗的传承现状/The Current Situation of Leno Inheritance

罗发展至今，江浙地区的罗呈现出明显的分化，浙江的杭罗、江苏的吴罗（四经绞罗和纱罗）分别成为两地的织造技艺代表。2008年，蚕丝织造技艺（杭罗织造技艺）入选国家级非物质文化遗产名录。2009年，蚕丝织造技艺（杭罗织造技艺）作为中国传统桑蚕丝织技艺的重要子项目之一被列入联合国教科文组织《人类非物质文化遗产代表作名录》。吴罗织造技艺（四经绞罗和纱罗）则入选第四批江苏省非物质文化遗产项目（表4-1）。

So far since leno has developed, leno in Jiangsu and Zhejiang regions has shown a clear differentiation, and Hang Leno in Zhejiang and Wu Leno in Jiangsu (four warp twisted leno and gauze) have become the representatives of weaving techniques in the two regions respectively. In 2008, the silk weaving technique (Hang Leno weaving technique) was selected in the National Intangible Cultural Heritage List, and in 2009, silk weaving technique (Hang Leno weaving technique), as one of the important sub-projects of traditional Chinese mulberry silk weaving skills, was inscribed as the *Representative List of the Intangible Cultural Heritage of Humanity* organized by UNESCO. Wu Leno was selected for the Fourth Batch of Provincial Intangible Cultural Heritage Projects in Jiangsu Province (Table 4-1).

表4-1　代表性的罗织造技艺

项目名称	入选名录	保护单位	传承人
蚕丝织造技艺（杭罗织造技艺）	2009年《人类非物质文化遗产代表作名录》	杭州福兴丝绸有限公司	邵官兴
蚕丝织造技艺（杭罗织造技艺）	2008年第二批国家级非物质文化遗产名录	杭州福兴丝绸有限公司	邵官兴
吴罗织造技艺（四经绞罗织造技艺）	2013年第六批苏州市非物质文化遗产代表性项目名录	吴中区苏州圣龙丝织绣品有限公司、吴中区光福镇文体教育服务中心、苏州市工业园区娄葑家明缂丝	李海龙、周家明

项目名称	入选名录	保护单位	传承人
吴罗织造技艺（纱罗织造技艺）	2016年第四批江苏省非物质文化遗产代表性项目名录	苏州市吴中区	李海龙
吴罗织造技艺（四经绞罗织造技艺）	2016年第四批江苏省非物质文化遗产代表性项目名录	苏州市工业园区	周家明

Table 4-1　Representative List of Leno Weaving Technique

Project	Lists Included	Unit of Protection	Inheritors
Silk Weaving Technigue（Hang Leno Weaving Technique）	Representative List of the Intangible Cultural Heritage of Humanity（2009）	Hangzhou Fuxing Silk Co. Ltd.	Shao Guanxing
Silk Weaving Technique（Hang Leno Weaving Technique）	National Intangible Cultural Heritage List（2008）	Hangzhou Fuxing Silk Co. Ltd.	Shao Guanxing
Wu Leno Weaving Technique（Four Warp Twisted Leno Weaving Technique）	The Sixth Batch of Representative Items of Intangible Cultural Heritage of Suzhou（2013）	Wuzhong District Suzhou Shenglong Silk Embroidery Co., Ltd, Wuzhong District Guangfu Town Cultural, Sports and Education Service Center, Suzhou Industrial Park Lou Turnip Jia Ming Woof	Li Hailong, Zhou Jiaming
Wu Leno Weaving Technique（Gauze Weaving Technique）	The Fourth Batch of Jiangsu Province Intangible Cultural Heritage Representative Projects（2016）	Wuzhong District, Suzhou	Li Hailong
Wu Leno Weaving Technique（Four Warp Twisted Leno Weaving Technique）	The Fourth Batch of Provincial Intangible Cultural Heritage Representative Projects（2016）	Suzhou Industrial Park	Zhou Jiaming

一、杭罗传承现状/Current Situation of Hang Leno Inheritance

杭罗与苏缎、云锦同为中国华东地区的三大丝绸名产。杭罗使用纯桑蚕丝以平纹和纱罗组织织造而成，其绸面有等距的直条形或横条形纱孔，孔眼清晰，质地柔软滑爽。这种丝织品穿着舒适凉爽，耐穿耐洗，多用作帐幔、夏季衬衫和便服面料等。

Hang Leno, Suzhou Satin and Yun Brocade are the three famous silk products in East China. Hang Leno is made of plain mulberry silk with plain weave and gauze. Its surface has equidistant straight or horizontal yarn holes, with clear holes and smooth texture. One could feel comfortable and cool when wearing this fabric. For it is durable and washable, Hang Leno is mostly used as curtains, summer shirts and casual clothes.

生产杭罗的织机几经变革，但其生产流程中仍保持着大量精细缜密的手工技艺，对生产者技能的要求极高。原料进厂后，必须经过严格检验、筛选，历经浸泡、晾干、翻丝、纤经、摇纤等系列工艺，然后才能上机织造。织成的粗坯还要经过精练、染色等工序，才能成为精致的杭罗。织机的一系列设置促使罗的织造从单纯的手工艺转变为半机械化的工程式操作，而传承人所传承的不仅是织造的手法，还有复杂的织造工艺设计和织机装造技艺，传承难度非常高，因而杭罗历来传承人不多。

The weaving machine used to produce Hang Leno has undergone several changes, but the production process still maintains a lot of fine and meticulous craftsmanship, requiring the highest level of skills of the producer. After the raw materials enter the factory, they must be strictly inspected and screened, going through a series of processes such as soaking, drying, turning, fibrillating and twisting before they can be woven on the machine. The rough woven one also needs to be refined and dyed before it turns to be delicate Hang Leno. A series of settings in loom has prompted the transformation of the leno weaving technique from a simple handicraft to a semi-mechanized engineering operation. Not only the weaving technique, but also the complex design of the weaving process and the loom's installation skills, which are difficult to be carried forward, thus there have not been many inheritors of Hang Leno.

织机的应用带来织造场地成本方面的问题。一台普通的绞经织机约2m宽、3m长，而且络丝、整经等工序均需专门的机器和场地。由此罗的织造成本提高，不仅有织机和丝线等硬性成本，还需织造工艺难度带来的附加成本、用人和用地等成本。因此，织造成本与收益的矛盾加剧了传承的难度。此外，在织造成本不断上升的同时，织造出的罗成品却没有相应地升值。

The presence of looms brings about problems in terms of room for weaving. An ordinary skein loom is about two meters wide and three meters long. Special machines and a large enough space are required for the winding and warping and other processes. The high cost of weaving leno involves not only rigid costs such as looms and silk threads, but also additional costs brought about by the difficulty of the weaving process, the cost of employment and space use. The contradiction between process cost and benefits exacerbates the difficulties of inheritance. In addition, while the cost of weaving had been rising, the value of the finished woven products had not been appreciated accordingly.

现代以来，罗逐渐从人们的视野中消失，掌握罗织造的技艺者也越来越少，大多数人已不知罗为何物。特别是在化学纤维织物的强烈冲击下，传统的丝绸生产举步维艰，杭罗生产尤为困难。目前生产杭罗的仅有杭州福兴丝绸厂一家，厂长出身于杭罗世家，直接传承了杭罗的织造技艺。杭州福兴丝绸厂生产的杭罗历来由北京"瑞蚨祥"、苏州"乾泰祥"等老字号经销，深受国内外消费者欢迎。杭罗生产正面临着重重困难，亟待保护。

Since modern times, leno has gradually disappeared from people's view, and fewer and fewer people have mastered leno weaving skills. Most people have no idea about what leno is. Especially under the strong impact of chemical fiber fabrics, traditional silk production is difficult, and Hang Leno production is particularly so. At present, only Hangzhou Fuxing Silk Factory produces

Hang Leno. The factory director comes from the family of Hang Leno, so he directly inherites the weaving skills of Hang Leno. Hang Leno produced by Hangzhou Fuxing Silk Factory has always been distributed by time–honored brands such as Beijing "Ruifuxiang" and Suzhou "Qiantaixiang", and is well received by domestic and foreign consumers. Now there is a lot of difficulties in the production of Hang Leno and is in urgent need of protection.

2005年9月，杭州福兴丝绸厂向政府有关部门报告，要求保护杭罗。杭州的杭罗织造技艺被发现后立即引起相关部门的重视，在各级政府的大力帮助和支持下，杭罗织造技艺现已先后列入杭州市级、浙江省级、国家级非物质文化遗产名录。2008年，杭罗同苏州缂丝、南京云锦等作为"中国蚕桑丝织"项目申报世界人类口头与非物质文化遗产。

In September 2005, Hangzhou Fuxing Silk Factory reported to the relevant government departments, requesting the protection of Hang Leno. Hangzhou's Hang Leno weaving technique attracted immediate attention of the relevant departments. With the great help of governments at all levels, Hang Leno weaving techniques have been listed in Hangzhou Municipal, Zhejiang Provincial and National Intangible Cultural Heritage List. In 2008, together with the Suzhou silk tapestry, Nanjing Yun Brocade and so on, Hang Leno was declared the world oral and intangible cultural heritage of humanity as the "Chinese Sericulture and Silk Weaving" project.

2009年9月30日在阿联酋首都阿布扎比召开的联合国教科文组织保护非物质文化遗产政府间委员会会议决定，"中国蚕桑丝织技艺"入选《人类非物质文化遗产代表作名录》。"杭罗织造技艺"是"中国蚕桑丝织技艺"中的重要代表性项目。

At the meeting of the UNESCO Intergovernmental Committee for the Protection of Intangible Cultural Heritage held in Abu Dhabi, United Arab Emirates on September 30, 2009, "Chinese Mulberry Silk Weaving Technique" was approved to be inscribed on the *Representative List of the Intangible Cultural Heritage of Humanity*. "Hang Leno Weaving Technique" is an important representative of "Chinese Mulberry Silk Weaving Technique".

二、吴罗（四经绞罗）传承现状/The Current Situation of Wu Leno (Four-warp Twisted Leno) Inheritance

古代吴地所产的罗称为吴罗，其中四经绞罗是吴罗的代表。四经绞罗是指以四根经丝为一组，左右相绞而形成较大孔眼的罗。商周时期，我国的织罗技术是以二经相绞的素罗为主，秦汉以后，出现了一种以二经相绞的链式绞组织，即将绞经轮流用左侧或右侧的地经交替，环环相扣，呈不分割的链状结构，分为以二经绞素罗和四经绞素罗以及用四经绞和二经绞交替起花的花罗，俗称四经绞罗（图4-1）。

In ancient times, the leno produced in the Wu region was called Wu Leno, of which the four–warp twisted leno was the most excellent. Four–warp twisted leno fabric is a leno weave with larger eyelets on which a pair of four–yarn threads are twisted. During the Shang and Zhou Dynasties, the weaving technology in China was mainly based on the two–warp twisted plain leno. After the

Qin and Han Dynasties, there appeared a kind of chain twisted structure with two strand warps, in which the strand warp alternated with the left or right side of the ground warp, and the loop was interlocked, and the chain structure was not divided. It is divided into two-warp twisted plain leno and four-warp twisted plain leno and the flower printed leno with four-warp twisted and two-warp twisted alternately, commonly known as four-warp twisted leno, as shown in Figure 4-1.

四经绞罗由于织造工艺相对复杂，织造时效率低下，因而四经绞罗在元末明初逐渐消失。清朝覆亡后，随着其主要消费群体的消亡和织造局的解体，大批量的吴罗织造随之停止。由于其织造难度大，制作成本高，加之受到洋布的冲击，吴罗技艺几近灭绝。

Due to its relatively complex manufacturing process and low efficiency during weaving, therefore four-warp twisted leno gradually disappeared in the late Yuan Dynasty and early Ming Dynasty. Especially at the end of the Qing Dynasty, with the demise of its main consumer group and the disintegration of the weaving bureau, the weaving of large quantities of Wu Leno then stopped. Because of the difficulty of weaving, the high cost of production and the impact of foreign cloth, the skills of Wu Leno were nearly extinct.

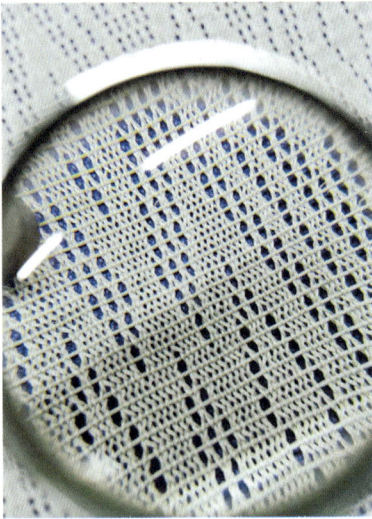

图4-1　放大镜下的四经绞罗
The Four-warp Twisted Leno under
the Magnifying Glass

吴罗织造技艺具有原真性的特点。由于光福镇地处苏州西部山区，过去交通不便，相对闭塞，且长期没有通电，致使流落在光福乡间的吴罗一直都只能靠脚踏手拉的老式织机生产，保持了它的原真性。

The Wu Leno weaving technique is characterized by its authenticity. Located in the mountainous area in the west of Suzhou, Guangfu was relatively closed and the traffic was not convenient, and there was no electricity for a long time, so the Wu Leno in the countryside of Guangfu has been produced only by the old looms with foot pedal and hand pull, which has kept its original authenticity.

吴罗织造技艺具有系列性的特点。目前所掌握的品种有素罗、横罗、直罗、对应连股罗、四经链式罗、绞花罗、芝麻花罗、实地花罗、亮地花罗、纹罗、妆花罗、绣花罗、挖花罗、漆纱罗、镂金罗等。

The Wu Leno weaving technique is characterized by its many varieties. At present, we have a variety of products, such as plain leno, horizontal leno, veritcal leno, and corresponding continuous strands leno, four-end link leno, twisted patterned leno, patterned flower with seamless dots, field flower leno, patterned leno with bright flower, and patterned leno, Zhuang Hua Leno, and embroidery leno, engraved flower leno, black jacquard leno, and engraved gold leno.

总之，吴罗织造技艺具有极高的历史文化价值、艺术审美价值和经济开发价值。因此，对吴罗加以传承和保护意义重大。

In conclusion, the Wu Leno weaving technique has great historical and cultural value, artistic and aesthetic value, as well as economic development value. Therefore, it is of great significance to inherit and protect Wu Leno.

第二节　罗织造技艺传承人/The Inheritors of Leno Weaving Technique

一、杭罗织造技艺传承人/Inheritors of Hang Leno Weaving Technique

邵官兴（图4-2），出生于1954年12月13日，男，汉族，浙江省杭州市福兴丝绸有限公司蚕丝织造技艺（杭罗织造技艺）传承人。入选第五批国家级非物质文化遗产代表性项目代表性传承人和浙江省第一批非物质文化遗产代表性传承人。

Shao Guanxing (Figure 4-2), born on December 13, 1954, male, Han nationality, is an inheritor of silk weaving technique (Hang Leno weaving technique) of Fuxing Silk Co. Ltd. in Hangzhou, Zhejiang Province. He was selected as the representative inheritor of the fifth batch of National Intangible Cultural Heritage Representative Projects and the first batch of representative inheritors of Intangible Cultural Heritage in Zhejiang Province.

图4-2　国家级非物质文化遗产传承人——邵官兴
Shao Guanxing, a National-level Intangible Cultural Heritage Inheritor

邵官兴出生在杭罗世家，他的祖辈从清光绪年间开始制作杭罗，爷爷邵明财在宣家埠小本经营杭罗，后来把这份产业和全套手艺传给了邵官兴的父亲邵锦全，传承到邵官兴时，已是第三代。邵官兴从小耳濡目染，从17岁开始，跟随父亲邵锦全学习织造杭罗。先学摇纡三年，再跟师父学习修理织机、整机，最后上机织造。在此期间，邵官兴还掌握了制作织机零部件的技术。后来邵官兴创办了杭州福兴丝绸厂，是目前世界唯一的杭罗生产地。

Shao Guanxing was born in a family that produced Hang Leno for generations. His ancestors started making Hang Leno during the Emperor Guangxu period of the Qing Dynasty. His grandfather, Shao Mingcai, ran a small business in Xuanjiabu, and later passed the industry and the full set of crafts to Shao Jinquan, Shao Guanxing's father. Shao Guanxin was already in the third generation by the time he inherited his father's business. Shao Guanxing grew up under the influence of his father's work, and at the age of 17, he followed his father to learn to weave Hang Leno. First, he learned to shake and twist for three years. Then he learned to repair the weaving pole and the whole machine with his master, and finally he got on the machine to weave. During this period, Shao Guanxing also learned to make the parts of loom. Later, he founded the Hangzhou Fuxing Silk Factory, which is currently the only place in the world to produce Hang Leno .

杭州福兴丝绸有限公司位于杭州九堡九昌路55号，这是一个远离闹市的僻静之地，正适合匠人潜心工艺。杭州福兴丝绸有限公司的前身是一间很简陋的厂房，由于国家政策的大力支持和资金资助，现在已经拥有3栋厂房和一个丝绸博物馆。

The Hangzhou Fuxing Silk Co. Ltd., located at No. 55 Jiuchang Road, Jiubao, Hangzhou, is a secluded place away from the busy streets, so it is suitable for craftsmen to concentrate on their craft. Hangzhou Fuxing Silk Co. Ltd. used to be a very simple and crude factory. It now has three factories and a silk museum thanks to the strong support both financially and in national policies.

图4-3　浙江省第二批非物质文化遗产
代表性传承人洪桂贞
Hong Guizhen, the Representative Inheritor of the Second Batch of Intangible Cultural Heritage in Zhejiang Province

洪桂贞（图4-3），出生于1956年8月9日，是杭罗传承人邵官兴的妻子，浙江省第二批非物质文化遗产代表性传承人。她从事杭罗织造已有三十年，对杭罗织机的构造了如指掌，特别是对杭罗织造的关键技术环节把握得恰到好处。

Hong Guizhen (Figure 4-3), born on August 9, 1956, is the wife of Shao Guanxing, the Hang Leno's inheritor, and the representative inheritor of the second batch of Intangible Cultural Heritage in Zhejiang Province. She has been engaged in Hang Leno weaving for thirty years, and knows the structure of Hang Leno loom well, especially the key technical aspects of Hang Leno weaving.

二、吴罗织造技艺传承人/Inheritors of Wu Leno Weaving Technique

李海龙（图4-4），出生于1952年，男，汉族，第五批江苏省非物质文化遗产项目吴罗织造技艺（纱罗织造技艺）代表性传承人，第四批苏州市非物质文化遗产项目吴罗织造技艺代表性传承人，中国社会科学院、考古研究所特聘研究员、中国文物学会纺织专业委员会理事、北京故宫乾隆花园修复专家、北京服装学院校外导师、江苏省高级工艺美术师。

Li Hailong (Figure 4-4), born in 1952, male, Han nationality, is the representative inheritor of Wu Leno weaving technique（leno weaving technique）, which was inscribed on the fifth batch of Jiangsu Provincial Intangible Cultural Heritage Projects and the fourth batch of Municipal Intangible Cultural Heritage Projects of Suzhou. He is the special research fellow of the Chinese Academy of Social Sciences and the Institute of Archaeology, director of the Textile Professional Committee of the Chinese Cultural Relics Society, and an expert in the restoration of the Qianlong Garden of the Forbidden City in Beijing. He is also an off-campus tutor of Beijing Institute of Fashion Technology, and a senior fine arts artist of Jiangsu Province.

1994年，李海龙拜郁石鸣为师开始钻研织罗技艺，2003年，他创办苏州圣龙丝织绣品有限公司从事纱罗及古丝绸产品的生产和研发。他除了掌握横罗、直罗等一般纱罗的织造技艺外，还恢复了花罗（亮花罗、暗花罗）、四经链式罗、妆花罗等诸多古老珍稀的织罗技艺，为复活漆雕罗、镂金罗、绣花罗等复合艺术作出了贡献。他曾为故宫博物院复制乾隆花园漆雕镂金罗窗纱等古罗实用艺术品，受到学界与业界的好评。2010年，他在光福工艺文化城建立了"锦罗艺术馆"，免费收徒，义务传承推广吴罗织造技艺，定期举行交流活动，管内陈列展示各种罗织物产品，免费对公众开放。

In 1994, thanks to his teacher, Yu Shiming, Li Hailong started to study the art of leno weaving. In 2003, he founded Suzhou Shenglong Silk Weaving & Embroidery Co. Ltd., which engaged in the production and research and development of leno and ancient silk products. In addition to mastering general leno weaving technique such as horizontal and vertical leno, he has also restored many ancient and rare leno weaving techniques, such as flower printed leno, including bright flower printed leno, dark flower printed leno, four-end link leno, Zhuang Hua Leno. He has also contributed to the revival of composite arts such as lacquer carved leno, engraved flower leno, and embroidery leno. He once reproduced the ancient lacquer-carved gold curtains in Qianlong's Garden and other practical works of art for the National Palace Museum, which have been well received by the academic and the industry communities. In 2010, he established the "Yun Brocade

图4-4　吴罗织造技艺（纱罗织造技艺）代表性传承人李海龙
Li Hailong, the Representative Inheritor of Wu Leno Weaving Technique（Leno Weaving Technique）

and Leno Art Museum" in Guangfu's Arts and Crafts Cultural City, which accepts apprentices for free and promotes the weaving skills of Wu Leno voluntarily, and would hold regular exchange activities and displayed various products of leno fabrics in the museum, which is open to the public for free.

周家明（图4-5），出生于1955年，男，汉族，第五批江苏省非物质文化遗产项目吴罗织造技艺（四经绞罗织造技艺）代表性传承人，第四批苏州市非物质文化遗产项目四经绞罗织造技艺代表性传承人。

Zhou Jiaming (Figure 4-5), born in 1955, male, Han nationality, is the representative inheritor of Wu Leno weaving technique (four-warp twisted leno weaving technique) in the fifth batch of Provincial Intangible Cultural Heritage Projects, and the representative inheritor of the four-warp twisted leno weaving technique in the fourth batch of Municipal Intangible Cultural Heritage Projects in Suzhou.

1982～1990年，周家明在苏州市漳绒丝织厂师从父亲织造丝绸，1990年创办苏州工业园区家明织造坊，从事缂丝、宋锦、漳缎及四经绞罗等织造，2012年至今，他用宋锦织造技艺为苏州丝绸博物馆小批量复制生产手工传统宋锦；2018年，他成功为南京云锦研究所复制了马王堆四经绞罗文物的样品。周家明熟练掌握花本制作、调丝、牵经、通经、摇纤、打平纹综泛头、打纹综泛头、打绞（脚）综、经线穿综等织造四经绞罗的工序。其制作的四经绞罗采用无筘织造技术，组织结构稳定，织物呈现轻、薄、透等特点，独特的绞综织造技艺，是有别于常规丝绸织造技艺的主要特征。

From 1982 to 1990, Zhou Jiaming learned silk weaving from his father in Suzhou Zhangzhou Velvet Factory. In 1990, he founded Jiaming Weaving Workshop in Suzhou Industrial Park, which engaged in woof, Song Brocade, Zhang Satin and four-warp twisted leno weaving and so on. Since 2012, he has reproduced small batches of handmade traditional Song Brocade for the Suzhou Silk Museum by Song Brocade weaving technique. In 2018, Zhou Jiaming successfully replicated the

图4-5　吴罗织造技艺（四经绞罗织造技艺）代表性传承人周家明
Zhou Jiaming, the Representative Inheritor of the Wu Leno Weaving Technique (Four-warp Twisted Leno Weaving Technique)

samples for reproducing the Mawangdui four-warp cultural relics for the Nanjing Yun Brocade Research Institute. Zhou Jiaming has mastered the process of weaving the four-warp twisted leno, such as making the pattern draft, adjusting silk, drawing warp, passing warp, shaking twist, beating the plain healds, beating the twisted healds, and threading the warp healds. This four-warp twisted leno adopts reed-free weaving technology, which has the characteristics of stable weave structure, light weight, thinness and transparency. The unique twisted heddle weaving technology is the main feature different from the conventional silk weaving technology.

朱立群（图4-6），苏州市第五批非物质文化遗产项目吴罗织造技艺代表性传承人，19岁开始从事纱罗织造。40多年的时间，他恢复了素罗、花罗、春罗、三经绞罗、四经绞罗、五经绞罗、妆花罗等珍贵丝织品种的织造技艺，并在2016年带着纱罗修缮了故宫的符望阁。

Zhu Liqun（Figure 4-6）, the representative inheritor of Suzhou's Intangible Cultural heritage Projects（Wu Leno weaving technique）, started to weave leno at the age of 19. Over the past 40 years, he restored the weaving technique of precious silk varieties such as plain leno, pattern leno, Chun leno, three-warp, four-warp and five-warp twisted leno, and Zhuang Hua Leno. In 2016, he also repaired the Forbidden City's Fuwang Pavilion by using leno and gauze.

图4-6　吴罗织造技艺代表性传承人朱立群
Zhu Liqun, the Representative Inheritor of Wu Leno Weaving Technique

第三节　罗传承与发展面临的困境/The Difficulties in Leno Inheritance and Development

一、代表性传承人老龄化严重，缺乏接班人/Aging of Representative Inheritors and Lack of Successors

目前，丝绸罗非物质文化遗产代表性传承人周家明、李海龙、朱立群等都已60多岁，虽然他们在企业的生产实践中培养了一些徒弟与技术人员，但是真正能熟练掌握罗织造原理和技能、能独立工作的人很少。代表性传承人老龄化严重，所以急需一批年轻人接班，充实罗织造技艺传承队伍。

At present, Zhou Jiaming, Li Hailong and Zhu Liqun, representative inheritors of leno, the intangible heritage, are all over 60 years old. Although they have trained some apprentices and technicians in the practice and production of the enterprise, few people can really grasp the principles and skills of leno weaving and has ability to work independently. The representative inheritors are aging seriously, and there is an urgent need for a group of young people to take over and enrich the inheritance team of leno weaving technique.

二、缺乏色彩、图案设计人员，制约罗新产品开发/Lack of Color and Pattern Designers, Restricting the Development of New Leno Products

随着时代的变迁，传统的丝绸图案和样式可能已经不符合大众审美和需求，因此要在确保传统技艺的基础上进行非遗产品的创新设计与发展，满足不同消费群体的需要。目前掌握丝绸罗传统织造技艺的人大多是从企业的生产实践中成长起来的，他们专注于某些工序的技术，而不懂产品的色彩搭配与织物图案设计，并且这些企业中也没有专门的图案设计人员，限制了丝绸罗非遗新产品的开发与创新设计。

As time goes on, the ancient traditional silk patterns and styles may no longer meet the aesthetic needs of the public. Therefore, innovative design and development of intangible heritage products should be carried out on the basis of ensuring traditional skills to meet the needs of different consumer groups. At present, most people who know the traditional weaving skills of leno grow up from the production of enterprises, focusing on the technology of certain processes, not knowing the color matching of products and fabric pattern design. Moreover, there are no special pattern designers in these enterprises, which limits the development and innovative design of new leno intangible products.

三、织造工艺复杂，效率低，生产成本高/Complex Weaving Process, Low-efficiency and High Production Costs

丝绸罗相对于其他织物，存在着绞经部分，增加了绞综，并且绞经相对于地经纱线的来回相绞，增加了织造时的工艺难度。在苏州工业园区家明织造坊，1个工人1天仅可织出40cm的罗织物面料，并且幅宽也不到50cm，即使在采用有梭织机的生产企业里，其织机车速也只有80~110r/min，生产效率仍然比较低，人工成本高。由于绞经时，开口机构要承受较大的张力，导致目前所能生产的丝罗织物的幅宽最宽也仅为1.28m。另外，近年来年轻人都不愿意到传统纺织企业工作，企业招工越来越难，用工成本越来越高。

Compared with other fabrics, there are strand warps in leno, which increases the twistedheald, besides, the strand warps move back and forth with the ground warp yarn, which increases the technological difficulty in weaving process. In Jiaming Weaving Workshop of Suzhou Industrial Park, one worker can only weave 40cm fabric a day with the width less than 50cm. Even in the production enterprises using shuttle looms, the speed of looms is only 80–110r/min, and the production efficiency is still relatively low with high labor costs. Because the opening mechanism has to bear a large tension when twisting warp, the widest width of the silk fabric that can be produced at present is only 1.28m. In addition, in recent years, young workers are reluctant to work in traditional textile enterprises, which makes it more and more difficult to recruit workers and makes the labor cost of enterprises increasingly higher .

四、产品出口量减少，国内消费趋于小众市场/Export Trades Decline while Domestic Market Niche

苏州企业生产的丝绸罗织物早期主要出口到日本、韩国，用来制作传统服装，近年来随着经济下滑，人民币升值，企业产品订单数量越来越少，有时一个样只有几米的订单。对于企业来说，一个产品花型就要制作一套花本，产量越低，意味着成本越高，企业利润越低。而国内市场，除了帮故宫博物院、丝绸博物馆等单位仿制与加工古丝绸产品外，新开发的丝巾、领带、衣服等产品在市场上销量不大，目前受众市场较小，有待开拓。

Silk and leno fabrics produced by Suzhou enterprises previously were mainly exported to Japan and the republic of Korea for traditional costumes. In recent years, with the economic slowdown and the appreciation of the RMB, the number of orders for products of enterprise has become smaller. Sometimes, they could only receive orders with a sample of only a few meters. For enterprises, it is necessary to make a set of pattern drafts for a product pattern. The lower output means the higher cost and lower profits of the enterprises. While in the domestic market, in addition to helping the Forbidden City, silk museum and other units to reproduce and process the ancient silk products, the enterprises newly developed silk scarves, ties, clothes and other products, but the sales in the domestic market are small. The current market is small and needs to be further explored.

第四节	罗的保护、传承与发展建议/Proposals for Leno Protection, Inheritance and Development

一、建立罗织造技艺的非遗生态圈/Establishing an Intangible Cultural Heritage Ecosystem for Leno Weaving Technique

非物质文化遗产的保护真正需要的是活态传承，因此整体的、群体性的、能够自然传承的保护更重要。江浙地区纺织相关的非物质文化遗产数量多、类型广，从蚕丝处理到织物后染整再到服饰民俗，形成了天然的纺织文化圈。国家级纺织相关非遗项目中，江苏省共9

项、浙江省共12项，见表4-2。

What the protection of intangible cultural heritage really needs is the live transmission, so the protection of the whole, the group, which can be passed on naturally, is more essential. There are many intangible cultural heritages related to textiles in Jiangsu and Zhejiang Provinces, covering from silk treatment, post-dyeing and finishing of the fabrics to costume folk customs, forming a natural textile cultural circle. As for the textile-related Intangible Cultural Heritage Projects in Jiangsu and Zhejiang Provinces, there are 9 projects in Jiangsu Province and 12 projects in Zhejiang Province, as shown in Table 4-2.

表4-2　江苏省和浙江省的国家级纺织相关非遗项目

省份	国家级纺织相关非遗项目名称	批次	区域
江苏	南京云锦木机妆花手工织造技艺	第一批	南京市
	宋锦织造技艺	第一批	苏州市
	苏州缂丝织造技艺	第一批	苏州市
	南通蓝印花布印染技艺	第一批	南通市
	金银细工制作技艺（江都金银细工制作技艺）	第二批	扬州市江都区
	金银细工制作技艺（南京宝庆银楼）	第二批	南京市
	南京云锦木机妆花手工织造技艺 扩展	第三批	南京市
	传统棉纺织技艺扩展（南通色织土布技艺）	第三批	南通市
	地毯织造技艺扩展（如皋丝毯织造技艺）	第五批	南通市如皋市
浙江	红帮裁缝技艺	第五批	宁波市奉化区
	花边制作技艺（萧山花边制作技艺）	第五批	杭州市萧山区
	彩带编织技艺（畲族彩带编织技艺）	第五批	丽水市景宁畲族自治县
	蓝印花布印染技艺	第四批	嘉兴市桐乡市
	中式服装制作技艺（振兴祥中式服装制作技艺）	第三批	杭州市上城区
	蚕丝织造技艺（杭州织锦技艺）	第三批	杭州市下城区
	蓝夹缬技艺	第三批	温州市
	蚕丝织造技艺（辑里湖丝手工制作技艺）	第三批	湖州市南浔区
	传统棉纺织技艺（余姚土布制作技艺）	第三批	宁波市余姚市
	蚕丝织造技艺（杭罗织造技艺）	第二批	杭州市江干区
	蚕丝织造技艺（余杭清水丝绵制作技艺）	第二批	杭州市余杭区
	蚕丝织造技艺（双林绫绢织造技艺）	第二批	湖州市南浔区

Table 4-2　National Textile-related Non-Foreign Heritage Projects
in Jiangsu and Zhejiang Provinces

Province	National-level textile-related Intangible Cultural Heritage Projects name	Batch	Region
Jiangsu	Nanjing Yun Brocade Wooden Machine Hand Weaving Technique	First Batch	Nanjing
	Song Brocade Weaving Technique	First Batch	Suzhou
	Suzhou Silk Tapestry Weaving Technique	First Batch	Suzhou
	Nantong Blue Printed Fabric Printing and Dyeing Technique	First Batch	Nantong
	Gold and Silver Fine-working Technique (Jiangdu Gold and Silver Fine-working Technique)	Second Batch	Jiangdu District, Yangzhou
	Gold and Silver Fine Art Production Technique (Nanjing Baoqing Silver House)	Second Batch	Nanjing
	Nanjing Yun Brocade Hand Weaving Technique on Wooden Machines Expanded	Third Batch	Nanjing
	Traditional Cotton Weaving Technique Expanded (Nantong Color Weaving Toile Technique)	Third Batch	Nantong
	Carpet Weaving Technique Extension (Rugao Silk Carpet Weaving Technique)	Fifth Batch	Rugao City, Nantong
Zhejiang	Red Gang Tailoring Skills	Fifth Batch	Fenghua District, Ningbo
	Lace Making Technique (Xiaoshan Lace Making Technique)	Fifth Batch	Xiaoshan District, Hangzhou
	Colored Ribbon Weaving Technique (She Colored Ribbon Weaving Technique)	Fifth Batch	Lishui, Jingning, She Autonomous County
	Blue Cloth with Design in White Dye-printing Technique	Fourth Batch	Tongxiang City, Jiaxing
	Chinese Clothing Making Technique (Zhenxing Xiang Chinese Clothing Making Technique)	Third Batch	Shangcheng District, Hangzhou
	Silk Weaving Technique (Hangzhou Brocade Weaving Technique)	Third Batch	Downtown, Hangzhou
	Blue Clad Valance Technique	Third Batch	Wenzhou
	Silk Weaving Technique (Serie Lake Silk Handcrafting Technique)	Third Batch	Nanxun District, Huzhou
	Traditional Cotton Weaving Technique (Yuyao Tufu Making Technique)	Third Batch	Yuyao City, Ningbo
	Silk Weaving Technique (Hang Leno Weaving Technique)	Second Batch	Jianggan District, Hangzhou
	Silk Weaving Technique (Yuhang Qing Shui Silk Mian Making Technique)	Second Batch	Yuhang District, Hangzhou
	Silk Weaving Technique (Shuanglin Damask Silk Weaving Technique)	Second Batch	Nanxun District, Huzhou

借助江浙地区纺织类非遗基数大的优势，建立纺织非遗生态圈，改变以往侧重于单个项目的保护机制，将优秀非遗纺织技艺和产品加以整合，将有紧密关联的传承内容加以联结，形成纺织类非物质文化遗产的资源链，打造整体和可持续的保护生态圈，从而为非遗传承提供生存环境、生活方式、风土人情的立体空间，使它们具有更加持续稳固的生命力。

With the advantage of the large base of textile Intangible Cultural Heritage in Jiangsu and Zhejiang Provinces, we can establish the textile Intangible Cultural Heritage ecosystem and change the previous protection mechanism that focused on individual items by integrating excellent Intangible Cultural Heritage textile techniques and products, forming a resource chain of textile Intangible Cultural Heritage by connecting closely related heritage contents, and creating an overall and sustainable protection ecosystem, thus providing a living environment, lifestyle, and customs for the inheritance of intangible cultural heritage, and make them more sustainable and stable.

从江浙地区现有的罗非遗传承状况来看，政府对传承项目和传承人的扶助与奖励政策还有待加强。在对现有传承人的补贴基础上，加大对下一代传承人的资助力度，鼓励青年和相关专业背景的学生拜师学艺，如可结合浙江理工大学"织锦传承人研习班"、中国丝绸博物馆"女红传习所"等教育平台，加强下一代纺织类传承人之间的联系，形成纺织技艺文化的人脉圈。

From the existing situation of leno inheritance in Jiangsu and Zhejiang Provinces, the government support and incentive policies for heritage projects and inheritors have yet to be strengthened. On the basis of subsidies for existing inheritors, we need to increase the financial support for the next generations of inheritors, encourage young people and students with relevant professional backgrounds to learn from their teachers. Cooperating with "Workshop for Brocade Inheritors" of Zhejiang Sci-Tech University , "Needle work Heritage Institute" of the China Silk Museum and other education platforms so as to strengthen the contact of the next generation of textile heritage, thus forming a social network for of textile skills and culture.

二、完善罗织造技艺的保护传承发展平台/Improving the Protection and Development Platform of Leno Weaving Technique

习近平总书记在党的十九大报告中提出，要坚定文化自信，推动社会主义文化繁荣兴盛。传统织物文化的再繁荣，依托于保护和宣传方式。完善历史技艺传承的线下展示，加大保护传承基地、博物馆和档案馆等的建设力度，利用非遗博物馆和非遗基地"讲好中国故事"。

In the report of the 19th National Congress of the Communist Party of China, General Secretary Xi Jinping proposed that cultural self-confidence should be firmly established and socialist culture should be promoted to prosperity. The re-prosperity of traditional fabric culture relies on protection and promotion methods. We will improve the offline display of the historical skill inheritance, increase the construction of the protection inheritance bases, museums and archives, and use the intangible cultural heritage museums and bases to "tell the Chinese story well".

浙江省杭州市江干区与杭罗保护单位杭州福兴丝绸有限公司通力合作，在杭罗的保护、传承、弘扬方面做了一系列的工作，包括建立保护传承基地、建设杭罗博物馆、完备杭罗档案等，值得大力推广和学习。

Jianggan District, Hangzhou City, Zhejiang Province and Hang Leno conservation unit Fuxing Silk Co., Ltd. have cooperated in the protection, inheritance and promotion of Hang Leno including the establishment of bases for the protection of heritage, the construction of Hang Leno Museum, and completion of Hang Leno archives and so on, which sets a good example for the conservation and promotion of Hang Leno.

杭州福兴丝绸有限公司建立了杭罗织造技艺的生产性保护基地，建成有2700m²的杭罗生产车间，有传统手工织机6台，半自动杭罗织机6台，传统牵经机1台，传统摇纡车2台，传统翻丝车1台。邵官兴、洪桂贞两位代表性传承人常年开展传习，教授学徒，培养技工，全厂23名技师在他们的带领下可熟练生产杭罗，为杭罗织造技艺的传承奠定基础。2016年杭州福兴丝绸有限公司获评第二批浙江省非遗生产性保护基地。

Hangzhou Fuxing Silk Co., Ltd. established a production base for the protection of Hang Leno weaving technique, and built a 2,700-square-meter Hang Leno production workshop, with six traditional hand looms, six semi-automatic Hang Leno looms, a traditional warp drawer, two traditional twisting carts, and a traditional silk turning carts. Shao Guanxing and Hong Guizhen, two representative inheritors, have been carrying out the training all year round, accepting apprentices and training technicians. 23 technicians in the factory all have good command of the production of Hang Leno under their leadership, laying the foundation for the inheritance of Hang Leno weaving technique. In 2016, it was awarded the second batch of Zhejiang Province Productive Protection Base of Intangible Cultural Heritage.

江干区和杭州福兴丝绸有限公司先后投资近200万元，在全社会和专业博物馆中征集各类有关杭罗的藏品，并专门在杭州福兴丝绸有限公司内建立了近1000m²的杭罗博物馆，展出藏品100余件，有杭罗的老式织机、织造杭罗有关的器具、历史上杭罗的成品、养蚕的器具等，面向公众开放，以扩大杭罗知名度。

Jianggan District and Hangzhou Fuxing Silk Co. Ltd. have invested nearly 2 million yuan to collect various types of collections related to Hang Leno in the country and professional museums and established the Hang Leno Museum in the Hangzhou Fuxing Silk Co. Ltd. With an area of nearly 1,000 square meters, the museum exhibits a collection of more than 100 pieces of Hang Leno old looms, including relevant appliances to weave Hang Leno, Hang Leno's finished products in ancient times, silkworm breeding equipment and other collections. This museum is open to the public and aims to expand the popularity of Hang Leno.

杭罗的相关档案主要分成三部分：杭罗的历史记载，包括书籍、戏曲、曲艺等；杭罗的保护文字资料集，包括从2002年开始的保护历程及过程中具有保护意义的书籍、报刊、政府公文、企业展会等；杭罗的相关器具、服饰、零件等。

The archives related to Hang Leno are mainly divided into three types: firstly, historical records of Hang Leno, including books, operas and operatic arts and so on; secondly, the textual collection of Hang Leno conservation, including books, newspapers, government documents, corporate exhibitions that have conservation significance during the conservation history and process since 2002; thirdly, related appliances of Hang Leno, costumes, parts, etc.

除对历史文本、织物及织机等文物的静态展示外，还加强了对织造技艺的活态展示，如邀请织造技艺的传承人在展馆现场进行织造。将非遗项目与传统的文化活动相结合，丰富杭罗宣传推广形式。例如，杭州根据杭罗织造技艺改编成《杭罗灯彩》《杭罗情丝》等舞蹈，并在比赛、社区等多个展示平台巡演。《杭罗灯彩》获省舞蹈大赛金奖，《杭罗情丝》参加2016年浙江省首届非遗春晚演出。在社会上发放《江干文博》等宣传资料，对外普及杭罗知识。

In addition to the static display of historical texts, fabrics and relics such as looms, more emphasis should be placed on the live display of weaving techniques by inviting the inheritors to perform weaving on site at the exhibition hall, combining the intangible cultural heritage projects with traditional cultural activities to enrich the promotion and publicity forms of Hang Leno. For example, Hangzhou has adapted the Hang Leno weaving technique into dances such as *Hang Leno Lantern* and *Hang Leno Love Silk*, and toured in some competitions and communities and other performing platforms. *Hang Leno Lantern* won the gold medal in the provincial dance competition, and *Hang Leno Love Silk* participated in the performance of the first provincial intangible cultural heritage Spring Festival Gala in 2016. We also distribute promotional books such as *Jiangganelor Museology* in society and popularize Hang Leno to the outside world.

为了拓宽罗推广的交流空间，应当探索多样化的展示方式和文化载体，多方位展示交流。罗作为其织造技艺的产出物，有相当高的工艺价值和实用价值，应充分利用博览会、展销会所提供的展示平台，借助特色服饰、工艺品、纪念品等载体，扩大非物质文化遗产的社会影响力。

To broaden the communication space for leno promotion, we shall explore diversified display methods and cultural carriers, and display and carry out communication and exchanges in multiple ways. As the product of leno weaving technique, leno has considerable craftsmanship and practical value. We could make full use of the display platform provided by expositions and fairs, and expand the social influence of intangible cultural heritage by means of carriers such as characteristic costumes, handicrafts and souvenirs.

建立织造技艺展示的线上平台，加强非遗保护传承的数字化建设。目前，江浙地区非遗保护中心的网站仅对相关的织造技艺有一些简单的介绍，数字化平台的建设还有待丰富和完善。利用网络平台记录非遗项目、公开历史资料、展示织造技艺可节约成本且易于推广，结合虚拟现实、增强现实、裸眼3D等先进技术动态展示织造细节，更能加强受众的互动积极性。

We should establish an online platform for the display of weaving technique and strengthen the digital construction of intangible cultural heritage protection and inheritance. At present, the website of the Intangible Cultural Heritage Protection Center in Jiangsu and Zhejiang has only some simple introduction to the relevant weaving skills, and the construction of the digital platform has yet to be continued. Using the online platform to record intangible cultural heritage projects to disclose historical information and display weaving technique can save costs and be easily promoted. Combined with advanced technology such as virtual reality, augmented reality and naked-eye 3D, weaving details could be displayed vividly so as to strengthen the interaction with the audience.

三、丰富罗织造技艺的宣传推广形式/Enriching Forms of Publicity and Promotion of Leno Weaving Technique

罗的非遗传承不仅是技艺本身的传承，罗产品除了作为纪念品、艺术品外，仍有非常实际的日常用途。罗制成的服饰和家用装饰品在历史文献中有丰富的记载，如罗衫、罗裙、纱帽、纱幔等，贯穿在生活中的各个角落。

The intangible cultural heritage of leno is not only the inheritance of the technique itself, but its product of leno, which is still useful in daily life in addition to being a souvenir and artwork. Costumes and home accessories made of leno are richly recorded in historical literature, with leno shirts, leno skirts, leno curtains throughout people's life.

从市场推广方面，传统的绞经丝织物在化纤织物的冲击下渐趋萎缩，难以在主流的面料市场中立足。要再次弘扬罗所蕴含的织物文化，不能仅限于文化层面的推广，其产业化的开发和保护机制有必要同步进行。通过一系列国际峰会的宣传，特色的宋锦服装作为国礼一炮而红，带动了大众对"新中式"服装的追捧。罗作为传统中式服装的常用面料，有一定的发展前景。除了大众接受度较高的主流服装，汉服等小众文化需求量也有所上升。近年来高校社团和民间组织开始流行汉服风尚，年轻一代对中国古代服饰文化的探索和尝试，对古代优秀纺织面料的市场来说是一个利好消息。

From the market promotion perspective, traditional twisted silk fabrics tend to shrink and it is difficult to gain a foothold in the mainstream fabric market under the impact of chemical fiber. To promote the fabric culture contained in leno again, it can not be limited to the cultural level of promotion. Its industrial development and protection mechanism needs to proceed as well. Through a series of international summits, the special Song Brocade clothing as a national gift has become a hit, driving the public to the "new Chinese style" clothing. As a common fabric for traditional Chinese clothing, leno also has a certain prospect. Besides mainstream clothing with high public acceptance, the demand for niche cultures such as Hanfu also increased. In recent years, university associations and civil organizations have started to popularize Hanfu, and the younger generation's exploration and experimentation with ancient Chinese costume culture is good news for the ancient excellent textile fabrics market.

四、在地方高校中传承丝绸非遗技艺和文化，充实传承人队伍/Passing on Silk Intangible Cultural Technique and Culture in Local Institutions of Higher Learning to Enrich the Team of Inheritors

《关于推进职业院校民族文化传承与创新工作的意见》（教职成〔2013〕2号）和《江苏省非物质文化遗产保护条例》(2013第125号) 相关规定，鼓励和支持地方高校、职业院校将优秀的丝绸类非物质文化遗产项目纳入专业课程教学体系，让更多的人了解丝绸类非物质文化遗产，培养更多的丝织专业人才。因此，拥有纺织与艺术类专业的相关地方高校需要承担起相应的责任与义务，开设丝绸类相关专业或者将丝绸类非遗传统技艺和文化嵌入相近的专业课程教学体系中，将丝绸类非遗传统技艺和文化传承发扬下去。鼓励与支持更多的纺织与艺术类高校、职业院校毕业生到丝绸企业工作，从而充实丝绸罗传承人的队伍。

According to related regulations in *Opinions on Promotion the Inheritance and Innovation of National Culture in Vocational Colleges and Universities* (Send by Department of Vocational and Adult Education〔2013〕No. 2) and the *Jiangsu Province Intangible Cultural Heritage Protection Regulations* (2013 No. 125), the Ministry of Education, the Ministry of Culture, the State Ethnic Affairs Commission encourage and support local universities and vocational colleges to incorporate outstanding silk intangible cultural heritage projects into the professional curriculum system so as to let more people understand silk intangible cultural heritage and train more silk specialists. Therefore, local colleges and universities with textile and art majors need to take the responsibility and obligation to offer silk–related majors or embed skills and culture of silk intangible heritage traditional into the teaching system of similar majors, so that the skills and culture can be carried forward. Encouragement and supports are given to more graduates from textile and art colleges and vocational school graduates to work in silk enterprises so as to enrich the silk leno inheritance team.

五、搭建产学研合作平台，推动罗产品的开发与创新设计/Building a Platform for Industry-university-research Cooperation to Promote the Development and Innovative Design of Leno Products

企业是丝绸罗传承与发展的主体，也是研发创新活动的主体，但是当前企业对于罗织造理论、新型织造技术了解偏少，不利于罗传统技艺与产品的科技创新。因此，可由企业与高校、研究所进行合作，共同建设研发平台；利用高校资源，开设相关课程，以多种形式为企业人员培训相关知识；利用纺织、艺术类高校教师和学生的优势，为企业开发图案，创新产品设计，推动罗非遗产品的创新与发展。鼓励丝绸类非遗织物生产企业与更多的产品设计公司进行对接，打通产品设计与织物生产两道环节之间的隔膜，开发出更多能够适应市场需要的丝绸类非遗产品。

Enterprises are the main body of leno inheritance and research and development innovation activities. However, these enterprises have little understanding about the current weaving theories and new weaving technology, which is not conducive to the traditional technique and products of scientific and technological innovation. Therefore, the enterprise can cooperate with universities and research institutes to jointly build research and development platforms, and use university resources to offer relevant courses to train enterprise personnel in various forms. They can also take advantage of teachers and students of textile and art colleges to develop patterns and innovate product designs for enterprises to promote the innovation and development of leno products. Silk-based fabric production enterprises should be encouraged to dock with more product design companies, bridging the gap between the two links of product design and fabric production, to develop more silk-based intangible cultural heritage products that can adapt to market needs.

六、推动罗织造技艺的工艺研发/Promoting Research and Development in Leno Weaving Technique

将传统技艺与现代技术结合，进行非遗项目的传承与创新，实现技术现代化，才会有更多的人进入丝绸行业。目前，罗采用传统木织机或有梭织机织造，生产效率低，幅宽窄，而宋锦早已采用剑杆织机织造，其车速可达350r/min，幅宽近2.8m。因此，可对罗的织造设备进行更新，有必要研究新型织造方式，如采用电子提花开口机构，减轻工人劳动强度，提高织造效率。政府相关部门可设立专项资金，鼓励企业技术革新，改造织机设备。技术革新后，生产效率提高，也会吸引更多的人才进入丝绸行业。

We should combine traditional skills with modern technology to inherit and innovate intangible cultural heritage items to realize the modernization of technology. In doing so, more people will participate in the silk industry. At present, leno is woven by traditional wooden loom or shuttle loom, which has low production efficiency and narrow width, while Song Brocade has long been woven by rapier loom, which has a speed of 350r/min and a width of nearly 2.8 meters. Therefore, the weaving equipment of leno needs to be updated, and it is necessary to study new weaving methods, such as adopting electronic jacquard shedding machine to reduce the labor intensity of workers and improve the weaving efficiency. Relevant government departments can set up special funds to encourage enterprises to innovate technology and transform loom equipment. The combination of technological innovation and higher production efficiency will also attract more people to engage in the silk industry.

非遗传承既需要文化创新，也离不开科技创新。"创意＋科技"的创新思路成为非遗传承突破瓶颈的有益探索，能促进更多像罗这样的顶尖国匠精品走入今日寻常百姓家，让更多人感受到国粹之美，期待创新的罗织物能交织出更灿烂的经纬未来。

Intangible cultural heritage inheritance requires both cultural innovation and technological innovation. The innovative idea of "Creativity+Technology" has become a feasible exploration of intangible cultural heritage to break through the bottleneck. It can also make such top national craftsmen products as leno more accessible to ordinary people today, so that more and more people can feel the beauty of the national silk treasure, and we are looking forward to interweaving a brighter future for leno fabric in innovative ways.

第五章

罗的创新与应用
The Innovation and Application of Leno

罗在古代是贵族阶层的必备之物，罗织物的纹样广泛汲取了传统文化的营养，如折枝牡丹、凤穿牡丹、豹首纹、云气纹、复合杯菱纹等都能在瓷器、陶器等传统工艺美术品中找到相应的出处。现今，纹样融入不同地域的文化风貌，顺应了文化消费的创新浪潮，作为"丝绸之府"的苏州，其地域文化风貌理应成为罗织物纹样创新的源泉。

Leno was essential to the aristocracy in ancient times. The pattern of Leno fabric has widely absorbed the nutrition of traditional culture, like folded peony, phoenix flying through peony, pattern of leopard's head, symbols of auspicious clouds, as well as rhombus pattern, etc., all of which can be found in china, pottery and other traditional arts and crafts. Nowadays, patterns with cultural design and scenery of different regions conform to innovation trend of cultural consumption. Suzhou is also known as the home of silk, whose regional culture should be a source of innovation for leno fabric patterns.

第一节　苏州地域文化及其视觉化提炼/An Overview and Visual Refinement of Suzhou Regional Culture

苏州是一座拥有2500年历史的文化古城，吴地民众创造了极为丰富的物质财富与精神财富，丝绸之府、文萃之邦、工艺之市、园林之城等美誉共同呈现其独特而鲜明的地域文化风貌。

Suzhou is a city with a history of 2,500 years. Since then, the people of Wu region have created bountiful material and spiritual wealth. The reputation of "the home of silk" "the land of outstanding literature" "the city of crafts" and "the city of gardens" jointly shows its unique and distinct regional cultural features.

苏州地域文化细腻。鱼米之乡的精耕细作，苏帮菜肴的食不厌精，非遗技艺的工巧艺精，苏绣针法的缜密，桃花坞年画刻版的技艺繁复，缂丝雕琢般的立体效果，它们都形象诠释着苏州地域文化的细腻与精巧。

Suzhou has delicate regional culture. Owning to the diligent people of Wu region, delicious Sue cuisine, the exquisite intangible cultural heritage, compacted weaving methods of embroidery, as well as the intricate carving technique of Taohuawu New Year pictures with the three-dimensional effect of kossu carving, the delicacy and fineness of Suzhou's regional culture has been interpreted vividly.

苏州地域文化雅致。碧波的太湖孕育着秀美的风光，古往今来文人墨客聚集于此，诸多文化和艺术碰撞交织，滋养出雅致的艺术气质。人们在意境清雅的园林中惬意地品鉴着碧螺春，欣赏着吴门画派的淡雅、苏派三雕的尚雅、昆曲艺术的高雅，这些儒雅之气满满渗透着地域文化的风雅与秀丽。

The culture of Suzhou is very elegant. By the rippling Taihu Lake lies the beautiful scenery, knowledgeable men of all ages preferred to gather here, creating a composite of distinct culture and arts and making Suzhou surrounded by artistic and elegant atmosphere. People can savor Biluochun Tea comfortably in a beautiful garden, appreciate the elegance of the paint of Wumen faction, the Three Sculptures of the Su school, and Kunqu Opera, all reflecting the tastefulness of regional culture of Suzhou.

苏州地域文化灵动。唐朝诗人杜荀鹤曾写道："君到姑苏见，人家尽枕河"，此画面是苏州水文化的灵动写照。柔和的水气息彰显着苏州慢生活的舒缓与轻灵；粉墙黛瓦与小桥流水的水巷民居体现着错落有致的灵动韵味；飘逸的苏州丝绸和清丽的水乡服饰写意出灵动的清新气息；婉转缠绵的昆曲和柔缓轻清的评弹令人领略到灵动韵味。

The culture of Suzhou is lively. Du Xunhe, a poet of the Tang Dynasty, wrote that "when you arrive at Gusu, you will see that all the houses there were built by the river", which vividly portrays Suzhou's water culture. The moist air here, reflects the ease and slow life of people living here. The houses with white walls, black roof-tiles show the well-distributed dynamic charming; the flowing Suzhou silk and beautiful water town costumes have depicted a vivid and fresh atmosphere; melodious Kunqu Opera and soft Pingtan make people appreciate the dynamic charm.

提取苏州地域文化的可视化元素，可从苏州的传统文化资源如园林、昆曲、宋锦、刺绣、年画、扇艺、玉雕等提取其间的造型、图案、色彩。苏州园林中的亭台楼阁、路桥廊径、山水泉石、园林的布局、多变的窗花等典型元素都可细化。昆曲剧装细腻的纹样、桃花坞年画的装饰手法、苏州砖雕精致吉祥的图案都是充满浓郁苏州地域元素的视觉化题材，这些元素结合飘逸柔和的罗织物可演绎出产品古朴而高雅的美。色彩上，粉墙黛瓦、碧水青石的民居色调层次丰富又淡雅沉稳，"浓墨淡彩，写意江南"给人一种宁静淡泊的色彩美感。视觉化的提炼还需将有形的物质文化和无形的精神文化融汇起来，形意结合，苏州地域文化里形成的内涵积淀是创意设计时挖掘不尽的宝藏。

To collect the visual elements of Suzhou regional culture, we can extract its shape, patterns and colors from Suzhou traditional cultural resources such as gardens, Kunqu Opera, Song Brocade, embroidery, New Year paintings, fan art, jade carving. Typical elements, such as pavilions, bridges and corridors, landscapes, garden layout, and varied paper–cuttings for window decorations in Suzhou garden can be refined. The exquisite patterns of Kunqu Opera costumes, the decorative technique of Taohuawu New Year picture and the delicate and auspicious designs of Suzhou brick carvings are all visual themes of Suzhou regional elements. Combined with flowing and soft leno fabrics, they will present the quaint and elegant beauty of products. In terms of color, white walls and black tiles, blue water blue stone dwellings are rich in tone levels and quietly elegant and calm. There is a sentence describing Jiangnan as a Chinese painting with soft colors, leaving people with tranquil and enjoyable imagination. The visual refinement also needs the integration of tangible material culture and intangible spiritual culture. The connotation formed in Suzhou regional culture is an inexhaustible treasure for creative design.

第二节　苏州地域文化在罗织物设计中的创新应用/The Innovative Application of Suzhou Regional Culture in Leno Fabric Design

　　《闲梦·芳春》(图5-1)以粉黛青瓦、小桥流水为设计元素，太湖石、桂花树点缀其中，加之通过蝴蝶、云雀等动态的描绘，原本静态的画面多了一抹春色和生机。蜿蜒的河水巧妙地形成了画面亮色与暗面的分割，亮色多为园林景致，玲珑俊秀的假山湖石、精雕细琢的冰裂花窗；暗面则是用装饰手法描绘的美丽苏城，再配以点线面等抽象元素，这给图案带来生动性的同时也呈现了和谐的美感。画面采用粉绿和粉紫的对比关系，搭配雅致沉稳，适用于系列产品的延续开发。

Leisure Dreams of *Fragrant Spring* (Figure 5–1) takes white walls and black tiles, small bridges and flowing water as its design elements, dotted with Taihu stones and osmanthus trees. In addition, through dynamic depictions of butterflies and larks, the original static picture endows a touch of spring and vitality. The meandering river subtly forms the division between the bright and dark sides of the picture. The bright colors are mostly garden scenery, rockery and lake stones, and finely carved ice–cracked flower windows. The dark side is the beautiful Suzhou depicted by decorative technique, coupled with abstract

图5-1 《闲梦·芳春》
Leisure Dreams of *Fragrant Spring*

图 5-2 《窗·语》
Window of *Language*

elements such as dots, lines and surfaces, which bring vividness to the pattern and also form a harmonious aesthetic. The picture adopts the contrast between pink green and pink purple, which is elegant and tranquil and is suitable for the development of series of products.

《窗·语》（图 5-2）是意象风格作品。作品以苏州园林花窗的单元体为原始素材，将之结构化和图形化。通过现代色彩与古老纹样的反复延展形成特有的节奏感与韵律感。这类线与面的秩序体现了微观和整体的美感关系，创造出了新鲜的生命，将传统文化很好地融入现代结构设计中。作品更换不同的色彩可形成不同的视觉体验，衍生到家纺、服饰、文创等方面，非常利于市场推广。

Window of *Language* (Figure 5-2) is a work of imaginative style, which is structured and graphically based on the units of windows in Suzhou garden. Through the repetition of extending modern colors and ancient patterns, the unique sense of rhythm is formed. This kind of line and plane order reflects the aesthetic relationship between micro and macro, creates a fresh life, and well integrates traditional culture into modern composition design. Different colors of works can form different visual experience, derivative to home textiles, clothing, cultural creation and other aspects, which is conductive to market promotion.

第三节　桃花坞年画元素在罗织物设计中的创新应用 / The Innovative Application of Taohuawu New Year Pictures Elements in Leno Fabric Design

桃花坞年画是苏州的文化名片，具有丰富的审美意趣，其元素要在设计中得到合理应用就应从艺术创意角度进行提炼，既要提取符合年画神韵精髓的元素，又需兼顾消费者的审美情趣和生活习俗。根据需要有些繁复的纹样需摒弃，有的则可有选择地进行二次设计，使主体元素更为突出，在表达上显得更为精炼，在开发时融入现代美学理念，或移情表达或嫁接内涵，从而创新设计出具有时尚非遗风格特征的罗织物产品。

Taohuawu New Year pictures, as a famous cultural brand of Suzhou, are of rich aesthetic, whose elements should be refined from the perspective of artistic creativity if they are to be applied in design. We should not only extract elements in line with the essence of the New Year pictures, but also take into account consumers' aesthetic taste and customs. Some complicated patterns should be abandoned according to the needs, while others can be redesigned selectively, so that the main elements are more prominent and the expression is more refined. Modern aesthetic concepts should be embedded through transfer of expression or grafting of connotation, so as to creatively design leno fabric products with fashionable intangible heritage style.

"福禄寿喜"罗织物设计（图5-3）体现了吉祥的主题。依据传统色彩的寓意首先确定系列产品的四种基调色彩。"福"基色为赭土色，"土"为大地，有四季平安、风调雨顺、人人希冀的祈福纳祥之意；"禄"基色为紫色，紫气东来是祥运的征兆，意味着事业昌盛；"寿"基色为青蓝色，寿比南山不老松，象征着心胸开阔，生命绵长；"喜"基色为中国红，意味着喜庆临门、好事连连。

"Fu Lu Shou Xi" is a design with auspicious theme（Figure 5-3）. According to the connotations of traditional colors, four keynote colors of series products are determined first. The basic color of "Fu" is ochre soil color, which is the color of the earth, symbolizing pleasant weather and safety; the basic color of "Lu" is purple, which is the symbol of auspiciousness and prosperity in business; the basic color of "Shou" is cyan blue, which is the symbol of longevity and broad-mindedness; the basic color of "Xi" is Chinese red, which represents festive joy and good luck.

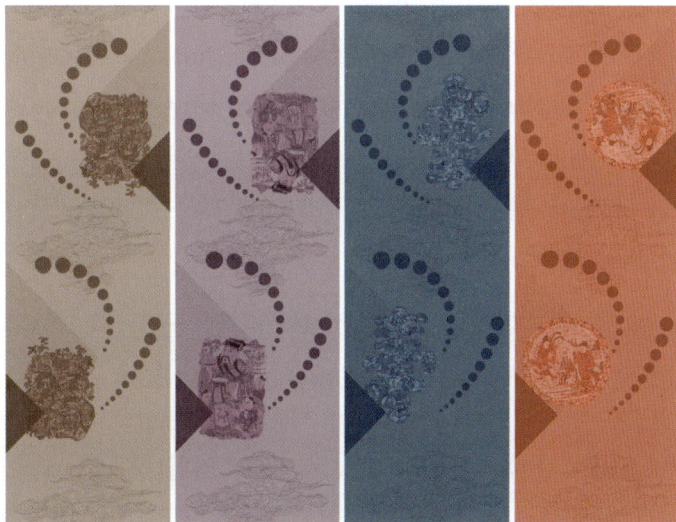

图5-3 "福禄寿喜"罗织物设计
Leno Fabric Pattern of "Fu Lu Shou Xi"

如图5-3所示，"福"单品采用桃花坞年画中的"福"字图案，综合了岁寒三友、刘海戏金蟾、和合二仙等吉祥纹样，再经色调加工润色完成。"禄"单品采用《天官赐福》的主体图形，福态的天官手捧仙鹿，传统文化中"鹿"谐音于"禄"，即为福禄之意。"寿"单品采用《寿字八仙》中的老寿星、八仙、寿桃、仙鹤等纹样合成，直观呈现了长寿之意。"喜"

单品采用《龙凤呈祥》的纹样，鸳鸯戏水、洞房花烛的欢乐情景呈现出永结同心的喜气氛围，寓意深受新婚夫妇的青睐。

As shown in Figure 5-3, the "Fu" item adopts the "Fu" pattern in Taohuawu New Year pictures, and also integrates the auspicious patterns such as the three companions of winter-pine trees, bamboo and yellow plum, Liuhai playing with gold toad, and fairy He-he, and then is finished by color processing and polishing. The "Lu" item adopts the main figure of *Heavenly God blesses the people*—the blessed official holding fairy deer in his hands. In traditional culture, "deer" is homophonic to "Lu", which means happiness and luck. The "Shou" item is mainly composed of the God of Longevity, eight immortals, birthday peach, crane and other patterns in *Shou Words with Eight Immortals* which intuitively presents the meaning of longevity. The "Xi" item adopts the pattern of *Dragon and Phoenix Bringing Auspiciousness*, and the joyful scene, like mandarin ducks playing in the water and joyfulness in wedding night, presenting a happy atmosphere of eternal unity, which is favored by newly weds.

除了主体的祥瑞图案，系列作品在构图中采用上下错位的设计，以线面流动的祥云纹样贯穿全篇，工整细腻的现代几何形点线面通过大小、反复、虚实的表现手法形成升腾的节奏感，活跃了画面构图的气韵与氛围。最终设计出的作品既呈现桃花坞年画的本真语境，又彰显出现代简约风的人文内涵。

In addition to the auspicious pattern of the main frame, this series of works adopt the up-and-down dislocation design in the composition, and the auspicious cloud pattern flowing through the whole work. The neat and delicate modern geometric points, lines and planes form a rising rhythm through the expression techniques of size, repetition and reality, which activates the charm and atmosphere of the picture composition. The final design work not only presents the true context of Taohuawu New Year pictures, but also highlights the humanistic connotation of modern simplicity.

| 第四节 | 昆曲元素在罗织物设计中的创新应用/The Innovative Application of Kunqu Opera Elements in Leno Fabric Design |

昆曲作为世界非物质文化遗产，其文化内涵与外在均有大量的可以挖掘利用的素材符号，可以从装饰艺术设计的角度，研究色彩与图形的现代审美表达，并将其应用于罗织物的图案设计中，从而达到提升产品文化价值的目的。

Kunqu Opera, as a worldwide intangible cultural heritage, has a large number of material symbols that can be elaborated and utilized in its cultural connotation and external presentation. From the perspective of decorative art design, we can study the modern aesthetic expression of colors and graphics, and apply them to the pattern design of leno fabric, so as to achieve the goal of enhancing the cultural value of products.

与传统的罗织物图案审美相比，现代的罗织物图案审美发生了巨大的变化，消费者关注的是产品整体的使用效果，而不是单纯地欣赏图案本身，图案设计的根本目标是满足人们对服饰品装饰美化功能的审美需求。因此，装饰图案的风格和构成样式应符合现代人的审美需求，尤以图案的抽象性表达最为重要。现代图形设计的特点是对抽象化重构手段的大量运用，图形符号的抽象化重构是运用抽象的视觉构成语言对图形重新进行艺术化诠释，是一种对具象元素进行再设计的手法，往往能够产生强烈的视觉冲击效果。昆曲视觉符号是一种抽象的概念和感受，必须通过设计方法将其转化成具体的纹样才能被更好地应用在罗织物设计中。具体来说，就是运用现代构成设计手法对传统视觉符号进行解构和重组。如图5-4所示，以昆曲脸谱为题材的图案设计中，打散脸谱本身对称的图形结构，经过解构和提炼，再以现代图案构成的方式进行重组，从而设计出具有脸谱意蕴的抽象纹样。

Compared with the traditional pattern aesthetics of leno fabric, the modern pattern aesthetics of leno fabric has undergone tremendous changes. Consumers pay more attention to the overall use effect of products rather than simply appreciate the pattern content itself. The fundamental goal of pattern design is to meet people's aesthetic needs for the decoration and beautification of clothing. Therefore, the style and composition of decorative patterns should meet the aesthetic needs of modern people, especially the abstract expression of patterns is of great importance. Modern graphic design is characterized by the extensive use of abstract reconstruction. The abstract reconstruction of graphic symbols is an artistic interpretation of graphics using abstract visual composition language, and it is a technique of redesigning concrete elements, which can often produce strong visual impact effect. The visual symbol of Kunqu Opera is an abstract concept and feeling, which must be transformed into concrete patterns through new design methods in order to be better applied to leno fabric design. Specifically, modern composition design techniques are to be used to deconstruct and reorganize traditional visual symbols. As shown in Figure 5-4, in the pattern design with Kunqu Opera facial makeup as the theme, the symmetrical graphic structure of facial makeup itself was broken up, deconstructed and refined, and then reorganized in the way of modern pattern composition, so as to design abstract patterns with facial makeup implication.

与传统的脸谱图案相比，打散重构后的图形更加自由、活泼、轻松、随意，同时兼具戏曲文化的要素特点，其文化意蕴的识别和感知并未受到影响。如图5-5所示，以昆曲生角的帽子为设计素材所进行的图形重构实践中，帽子的轮廓、装饰纹样等元素被分解、提炼，再利用重复和特异等构成手法设计出具有隐喻性

图5-4 昆曲脸谱罗织物设计
Leno Fabric Design of Kunqu Opera Facial Makeup

和指代性的抽象图形。在现代社会中，人们对织物装饰图案的纹样要求变得越来越趋于多样化和个性化。一目了然的图形不利于引发人们深入探究的兴趣，同时也缺乏审美互动的乐趣。因此从设计的角度来说，对罗织物图案纹样造型及整体构图方面的创意设计是非常关键的部分。

Compared with the traditional facial makeup patterns, the reconstructed patterns are more free, lively, relaxed and casual, and have the characteristics of opera culture, while the recognition and perception of its cultural implication have not been affected. As shown in Figure 5-5, in the practice of graphic reconstruction based on the Sheng's (main male role) hat in Kunqu Opera, the outline, decorative patterns and other elements of the hat are decomposed and refined, and then abstract graphics with metaphor and reference are designed by means of repetition and specificity. In modern society, people's requirements for fabric decorative patterns are becoming more and more diversified and personalized. Clear-cut graphics are not conducive to arousing in-depth interest and lack of fun in aesthetic interaction. Therefore, from design perspective, the creative design of leno fabric pattern modeling and overall composition is the key part.

图5-5　昆曲生角帽子罗织物设计
Leno Fabric Design of Sheng's Hat in Kunqu Opera

○ 参考文献
References

［1］朱新予. 中国丝绸史：专论［M］. 北京：中国纺织出版社，1997.

［2］袁宣萍. 浙罗［M］. 苏州：苏州大学出版社，2011.

［3］李海龙. 吴罗［M］. 南京：江苏凤凰教育出版社，2022.

［4］赵丰. 中国丝绸艺术史［M］. 北京：文物出版社，2005.

［5］《中国近代纺织史》编辑委员会. 中国近代纺织史：下卷［M］. 北京：中国纺织出版社，1997.

［6］吴县地方志编纂委员会. 吴县志［M］. 上海：上海古籍出版社，1994.

［7］罗群. 缥缈变化的"宋"罗［J］. 丝绸，2012，49（4）：52-56.

［8］宋应星.《天工开物》［M］. 台北：台北世界书局，1996.

［9］胡小鹏. 中国手工业经济通史：宋元卷［M］. 福州：福建人民出版社，2004.

［10］夏正兴. 中国古代罗织物［J］. 上海纺织工学院学报，1979（3）：86-92.

［11］罗群. 古代提花四经绞罗生产工艺探秘［J］. 文物保护与考古科学，2008（5）：20-25.

［12］郝鸿江. 杭罗织造技艺的起源和演变研究［D］. 上海：东华大学，2015.

［13］王子琪. 中国古代绞经织物结构、技艺与文化研究［D］. 上海：东华大学，2021.

［14］顾平. 织物组织与结构学［M］. 上海：东华大学出版社，2010.

［15］葛鸿鹄. 织物结构与性能［M］. 南京：江苏凤凰教育出版社，2015.

［16］郝鸿江. 杭罗织造技艺的起源和演变研究［D］. 上海：东华大学，2015.

［17］张迪，乐山. 机械时代传统工艺的消解与再生——以苏州四经绞罗织造技艺为例［J］. 艺术生活-福州大学学报（艺术版），2020（1）：20-28.

［18］丝绸文化与产品编写组. 代表性丝绸面料（7）：纱罗类面料［J］. 现代丝绸科学与技术，2019，34（2）：38-40.

［19］牛建涛，胡绮，黄紫娟，等. 苏州非物质文化遗产丝绸罗的保护与传承现状分析［J］. 丝绸，2018，55（10）：78-82.

［20］王子琪，邓可卉，张雷. 基于非遗调查的江浙地区绞经织物工艺传承与发展［J］. 丝绸，2018，55（10）：83-90.

［21］吕婷，卢业虎. 基于LoRa的服装陈列灯光控制系统设计［J］. 现代丝绸科学与技

术，2020，35（6）：24-27，34.

［22］邹启华，周青奇. 昆曲妆面视觉符号在纺织品图案设计中的应用［J］. 现代丝绸科学与技术，2020，35（6）：31-34.

［23］汤晓颖，谢瑅. 浅析《牡丹亭》旦角妆面的视觉符号意义及美学特征［J］. 包装世界，2013（6）：108-109.

［24］张杨. 中国传统视觉符号在平面设计中的应用研究［J］. 美术界，2013（1）：103.

［25］温润. 论现代家纺设计中流行色的运用［J］. 丝绸，2009（6）：16-19.

［26］邢庆华. 论后现代文化思潮下当代中国家用纺织品图案设计战略［J］. 南京艺术学院学报（美术与设计版），2004（4）：70-73.

［27］张琴. 蓝花布上的昆曲［M］. 北京：生活·读书·新知三联书店，2008.